Contents

Part 4: Illustrating the links

Part 5: Examination advice

Appendix

Population
& Migration

Michael Witherick

Series Editor: Sue Warn

Philip Allan Updates
Market Place
Deddington
Oxfordshire
OX15 0SE

Tel: 01869 338652
Fax: 01869 337590
e-mail: sales@philipallan.co.uk
www.philipallan.co.uk

ISBN-13: 978-1-84489-203-7
ISBN-10: 1-84489-203-4

Front cover photograph reproduced by permission of Ulrike Preuss/Photofusion

Printed by Raithby, Lawrence & Co. Ltd, Leicester

Environmental information
The paper on which this title is printed is sourced from managed, sustainable forests.

Introduction

Much of population geography is about the **distribution** of people on the Earth's surface and the reasons behind this. However, the emphasis is not only on how and why population numbers and densities vary. Population distribution is dynamic; it is changing constantly in response to variation over time in the rate of **natural population change**. Natural population change is the outcome of shifts in the balance between **fertility** and **mortality**, between **birth rates** and **death rates**. There are also changes in population structure, for example in age and sex ratios or in ethnic and household composition. These vary spatially too. Another powerful influence on the changing map of population distribution is **migration**. This is the movement of people over the face of the Earth, for example from inner city to suburb or from one continent or country to another.

The changes identified so far open up another field of population geography, one that focuses on the impact of people (particularly population growth) on the **consumption** of resources, goods and services. Levels of consumption vary from place to place and have a direct bearing on standards of living and human welfare.

Finally, population geography recognises that in today's world, the **policies** of governments, together with those of other decision makers and managers, try to influence various aspects of population — distribution, change, migration and consumption.

About this book

There are five main components of population geography:
- distribution
- change
- migration
- consumption
- policy

Distribution, change, migration and consumption are covered in the first two parts of this book, with the support of a range of up-to-date case studies. However, it would be wrong to think of these key ideas as freestanding components. They are interlinked, particularly through policy. The third section of the book examines policies and issues and the fourth explores the links between the components by means of more case studies. It concludes with two overarching studies, the asylum-seeker issue in the UK and AIDS in southern Africa. These serve to illustrate the point that there are interconnections between all five components (Figure 1).

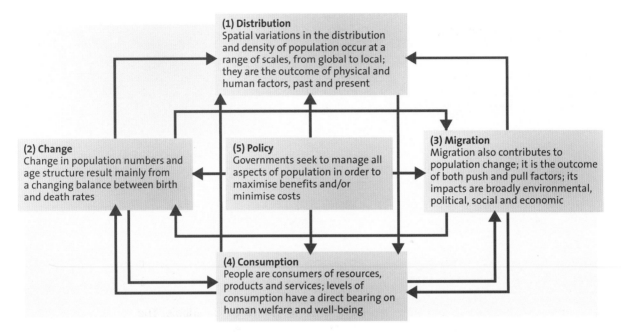

(1) Distribution
Spatial variations in the distribution and density of population occur at a range of scales, from global to local; they are the outcome of physical and human factors, past and present

(2) Change
Change in population numbers and age structure result mainly from a changing balance between birth and death rates

(5) Policy
Governments seek to manage all aspects of population in order to maximise benefits and/or minimise costs

(3) Migration
Migration also contributes to population change; it is the outcome of both push and pull factors; its impacts are broadly environmental, political, social and economic

(4) Consumption
People are consumers of resources, products and services; levels of consumption have a direct bearing on human welfare and well-being

Figure 1
The key ideas and components of population geography

Advice is given throughout on making the best use of case studies at both AS and A2. After some of the case studies, there are *Using case studies* boxes. Most of these show how a particular case study might be useful for answering a specific question; some invite you to try your hand at an exercise based on one or more of the case studies. You will also find some tips on tackling tasks that are commonly required in examinations, such as:

- describing the main features of a population distribution map
- identifying the basic character of a population pyramid
- extracting the message conveyed by a statistical table
- analysing a newspaper cutting

The final part of the book consolidates much of this advice and gives some additional tips on making the most of your case study material in the examination.

Key terms

Asylum seeker: a person who seeks to gain entry to another country by claiming to be a victim of persecution, hardship or some other compelling circumstance.

Birth rate: the number of live births per 1000 people in a year; an indirect measure of a population's fertility.

Carrying capacity: the maximum number of people that can be supported by the resources and technology of a given area.

Change: commonly taken to mean the percentage change in population numbers (i.e. growth or decline) over a year. Strictly speaking, the term includes other shifts over time, particularly in density, structure, migration and consumption.

Consumption: the resources, goods and services used by a population.

Death rate: the number of deaths per 1000 inhabitants of a given population in a year; a measure of mortality.

Density: the number of people per unit area (usually km^2).

Dependency ratio: the number of children (aged under 15) and old people (aged 65 and over) expressed as a ratio to the number of adults aged between 15 and 64. It indicates the number of people that the working population has to support.

Distribution: where people are located within a given area.

Ethnic cleansing: a euphemism for the actions of one ethnic or religious group forcing another such group to flee their homes, either by eviction or through fear and intimidation.

Fertility: the potential of a population to reproduce, often measured as the number of children (under 5 years) per woman of child-bearing age (15–50 years) or as the average number of children born per woman.

Human development index (HDI): a measure of national development, which can be used for making international comparisons. The index is based on three equally weighted variables: income per capita, adult literacy and life expectancy. The index takes the lowest and highest values recorded in the world for each variable. The interval between them is given a value of 1 and the figure recorded by each country is then scored on a scale of 0 to 1 (from worst to best). The HDI is the average score of the three variables.

Migration: the permanent or semi-permanent movement of people from one place to another. The balance between arrivals in, and departures from, an area impacts on change.

Mortality: the incidence of death in a population over a given period.

Natural change: the outcome of the difference between birth and death rates; growth results when births exceed deaths and decline occurs when deaths exceed births.

Optimum population: a theoretically perfect situation in which the population of an area can develop its resources to the best extent and enjoy the highest possible standard of living.

Overpopulation: a situation where the population of an area exceeds its carrying capacity. The symptoms are low (even declining) per capita income and standards of living, unemployment and outward migration.

Policies: the ways in which governments try to manage aspects of the population, for example change and distribution.

Refugee: defined by the UN as someone whose reasons for moving are genuinely to do with fear of persecution or death.

Remittances: money sent home to family members by migrants working and living abroad.

Structure: the make-up of a population analysed in terms of, for example, age, sex, marital status, ethnicity, family size and household characteristics.

Underpopulation: a situation where the resources and development of an area could support a larger population without any lowering of the standard of living *or* where a population is too small to develop its resources effectively.

Population distribution and change

The census

Figure 1 (page vi) shows the five components of population geography. Before looking at these in more detail and illustrating them by means of case studies, we need to consider the data that underpin most studies of population and migration. What information is needed and where does it come from?

Referring to Figure 1, reliable statistical information is needed about:

- total numbers of people and their distribution at a range of spatial scales from global to local (component 1)
- population structure, for example age, sex, marital status, ethnic origin and household size (components 1 and 2)
- population change, for example rate and scale as reflected in birth and death rates, or by the numbers of migrants entering and leaving an area (components 2 and 3)
- human welfare, as measured by a range of indicators, for example unemployment, per capita income and housing conditions (component 4)

Policy (component 5) is likely to need most of the above statistical information, and more besides.

The main source of such statistical data is the national census. Unfortunately, not all countries have the means to conduct a census at regular intervals because it is expensive and requires a high level of organisation. Even where census data are available, their quality and accuracy vary enormously from country to country. In general terms, it is the MEDCs that have the resources and organisational structures to collect large quantities of reliable information. However, the first part of the following case study of the UK's latest 10-yearly census (2001) offers a word of caution. So too does the remainder, but for different reasons.

THE UK'S 2001 CENSUS

Case study 1

Three cautionary tales

The missing million

The 2001 census put the total population of the UK at 58 789 194. Despite the legal requirement for every householder to complete a census form, inevitably there were a certain number of people who, for various reasons, did not do so. We are told that 94% of the census forms distributed across the country were filled in. However, in the ten boroughs of inner London, for example, the overall completion rate was below 80%; in Kensington and Chelsea it was only 64%. Four per cent of the national forms were subsequently 'estimated' on the basis of a follow-up survey of 300000 people. The remaining 2%, or about 1 million people, were 'invented' by census officials. Even so, the total population was still almost a million fewer than had previously been forecast by the Registrar General. Census officials believe that the shortfall is partly explained by young men disappearing into the Mediterranean rave culture and by students (mainly male) going around the world on gap-year trips. Apparently, 600000 more people than had been thought were out of the country. Border control figures collected from around the world suggest that 50000 were young UK citizens travelling in Australia.

The low response rate, plus the fact that the final population figure was 900000 fewer than had been predicted, has brought calls for 'a more robust and accurate method of measuring population'. But is there one?

This part of the case study illustrates the point that even in a well-organised MEDC, no matter the care taken, the results of a census are not always as complete or as accurate as might be thought.

Population growth: a red light

Despite the missing million, the census results show that the UK has the fastest-growing population of any large country in Europe. Its population is now predicted to increase from the present 59 million to 66 million by 2050. Having been stable for several decades, this take-off in population is ringing alarm bells. In response the government has commissioned a study to examine the impact on British life, particularly on the quality of life, infrastructure and the environment. This investigation could lead to the adoption of a population policy that tries to slow down the rate of increase.

The last time the government considered the issue of population growth was in 1973, when a study concluded that 'Britain would do better in future with a stationary rather than an increasing population'. The reasons given were that it would help relieve the problems of food supply and an unfavourable balance of payments. As it happened, the government took no action because a falling birth rate, together with low levels of immigration and high levels of emigration, ensured that the UK had a relatively stable population from the mid-1970s to the mid-1990s. However, the present Labour government's policy of encouraging immigration to Britain, in order to bridge a skills shortage, has re-ignited population growth, which is now at its highest rate since the 1970s. The population is increasing by about 250000 people a year, with about three-quarters accounted for by immigration. Also helping to raise the rate of population growth are the relatively high birth rates among some of the immigrant groups.

This part of the case study makes a number of key points:
- *Close monitoring of population trends is made possible by regular census taking.*
- *Government policy, along with any one of the three components of population change — birth rate, death rate and migration — can have dramatic consequences.*
- *One of the responsibilities of government is to ensure that the national population is kept as close as possible to an optimum level.*
- *How does the UK's recent experience fit the demographic transition model? Is the UK regressing to Stage 3 or is it defining a new Stage 5?*

Contemporary Case Studies

Nosy neighbours

A new government website providing information on everything from house prices to marital status means that homebuyers will be able to make detailed checks on the type of people likely to be their new neighbours. Information from the 2001 census has been broken down into more than 175 000 small areas, each comprising around 125 households. This has given the public free access to more personal information about their neighbours than ever before. In addition to data from the census, there is information from 25 other official sources, including data on house prices, land registry and crime.

By entering their postcode into the website **www.neighbourhood.statistics.gov.uk** and zooming into a street or estate, users can find out the racial and religious backgrounds of the inhabitants and check how many have university degrees. Homebuyers worried about whether it will be easy to find a parking space can check how many homes in a street have one, two, three or four cars and how many have none. They can also find out how many professional people live in the neighbourhood and how many house-holders have paid off their mortgages. Parents are able to find out whether children living in the street are the same age as theirs, while full-time mothers are able to check the number of women available for coffee mornings because they do not go out to work.

As the director of the Neighbourhood Statistics and Census Outputs has put it: 'the information contained in 50 gigabytes of on-line data "will tell thousands of stories that have never been told" '. At the same time, he has insisted that every effort has been made to ensure that no individual is identifiable.

Neighbourhood information is useful not only to homebuyers and local government officials. Other possible interest groups include estate agents, marketing companies, criminals and simply nosy people wishing to know a little more about their, and other, neighbourhoods.

This part of the case study gives:
- *a warning about data protection and the possible misuse of data*
- *some idea of the great wealth of data collected by the most recent UK census*
- *a very good website for accessing data that you might use in your personal enquiry*

Distribution

Awareness of the difference between distribution and **density** is vital to under-standing this component of population geography. Distribution is where people are located in an area. Population distribution is most simply shown on a map by repre-senting each person or group of people by a dot or symbol. This shows us where people are and gives an impression of how the numbers of people vary from place to place. One of the most useful indications of population distribution is achieved when population numbers are related to the space they occupy. This is density — the number of people per unit area (km^2 or hectare).

The most crucial fact of population geography is that people are not evenly distri-buted across the face of the Earth. This can be seen at a range of spatial scales from global to local. In 2000, the world's population was an estimated 6057 million, occupying the world's land area, including Antarctica, of 150 million km^2. This gives a mean population density figure of 40 persons per km^2. Over two-thirds of the global population live in Europe and Asia, but these two continents account for little more than one-third of the total land area.

While continental population densities range from 2.1 persons per km^2 in Australasia to 62.0 persons per km^2 in Asia, looking more closely at a population distribution map reveals far greater extremes. There are considerable variations at a national level. In Asia, for example, density ranges from nearly 6500 persons per km^2 in Singapore to a mere 2 persons per km^2 in Mongolia. The two case studies that follow illustrate that, as the spatial focus is narrowed, yet more density contrasts become evident, between regions and, most strikingly, between urban and rural areas.

The number of people in a given area, and their distribution within it, are the outcome of a complex interaction of factors, as shown in Figure 2. Present population should be thought of as the product of population trends operating over long periods of time. These population trends are not always upward. Populations can and do stagnate; they can even decline. In short, **population change** does not always mean population growth. As will be seen in the next section, change is the result of the overall and constantly shifting balance between births, deaths and migration.

Figure 2
Factors affecting the distribution of population

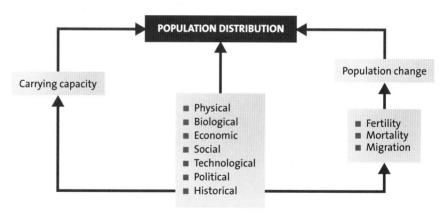

Figure 2 shows that population distribution is also influenced by another key factor, **carrying capacity**, namely the maximum number of people that can be supported by the resources and technology of a given area. However, there is a wide range of factors affecting the key influences on population distribution. These factors might be thought of as both creating opportunities and setting limitations (i.e. constraining).

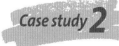

Case study 2

POPULATION DISTRIBUTION, CHANGE AND CARRYING CAPACITY

A look at Ethiopia

Ethiopia's population size and rate of growth are among the highest in Africa. In 2004, the population was estimated at around 67 million, making it the second most populous sub-Saharan African country after Nigeria. With an annual growth rate estimated at 2.9%, Ethiopia's population will approach 110 million before 2020 —

double what it was at the time of the last census in 1994. Nearly 2 million people are added to the country's population each year. This rapid population growth, the result of high fertility and a rapid lowering of the death rate, represents a serious obstacle to Ethiopia ever achieving sustainable development.

Over 75% of Ethiopians live in rural areas and a significant percentage of these people are pastoral nomads, which makes counting the population both difficult and inaccurate. With per capita GDP at only a little over $100, Ethiopia is one of the poorest countries in the world.

Looking at the distribution of population in terms of density, the national average is 55 persons per km². At a regional level, densities range from less than 30 to over 100 persons per km² (Figure 3).

The distribution pattern is strongly conditioned by altitude and relief. Physically, Ethiopia consists of two great plateaux separated by part of the Great Rift Valley (Figure 4).

The Ethiopian Plateau to the west of the Rift Valley is the most fertile and populated part of the country. The highest densities of population centre on the capital city, Addis Ababa (population 6 million). East of the Rift Valley is the Somali Plateau, which rises to over 4250 m in the Bale Mountains before sloping gently eastward to the Ogaden Plateau. The Rift Valley separating the two plateaux is a long, narrow cleft dotted with lakes that broadens to the north to form the Danakil Depression, an extensive desert.

The large range in altitude, from less than 500 m to over 4000 m, has a considerable impact on climate and it is climate that ultimately affects the distribution of population. About 15% of the population live in areas above 2400 m, around 75% live in the zone between 1500 and 2400 m, and only about 10% live below 1500 m — despite the fact that over half of Ethiopia's territory falls in this category. Locations above 3000 m and below 1500 m are sparsely populated. Above 3000 m the terrain is rugged and temperatures are low. These factors limit agriculture. Below 1500 m (apart from in the west and southwest) there are high temperatures, low rainfall and recurrent drought.

Strong though the influence of the physical environment is on the distribution of population, other

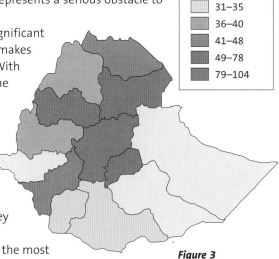

Persons per km²	
	8–30
	31–35
	36–40
	41–48
	49–78
	79–104

Figure 3
Ethiopia: population distribution, 2000

Figure 4
Ethiopia: relief and drainage

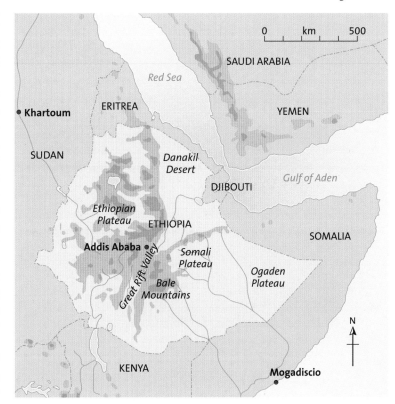

factors have been, and are still, at work. A number of non-physical factors indicated by Figure 2 are outlined below:

- **Biological factors:** rates of population growth are high in the most populated and more prosperous regions. Malaria and other chronic illnesses prevail in the lower parts of the country and act as a deterrent to settlement.
- **Economic factors:** the economy is based on agriculture; 75% of the population is dependent directly on farming and livestock rearing. Developing other means of livelihood might help to raise the national carrying capacity, but the opportunities are limited (tourism is one possibility). However, they are unlikely to make much impact on the overall distribution of population.
- **Social factors:** the pull of Addis Ababa is such that massive rural–urban migration is accentuating the concentration of population on the Ethiopian Plateau.
- **Technological factors:** because Ethiopia is such a poor country, the available technology can do little to open up the more remote areas and to improve the agricultural productivity of more marginal environments.
- **Political factors:** a longstanding border dispute with Eritrea (formerly part of Ethiopia) means that money that could be used to deal with the demographic situation is spent instead on the armed forces. Inhabitants have been forced to move away from the dangerous frontier zone.

Figure 5
Ethiopia: the level of need for emergency food assistance, 2003

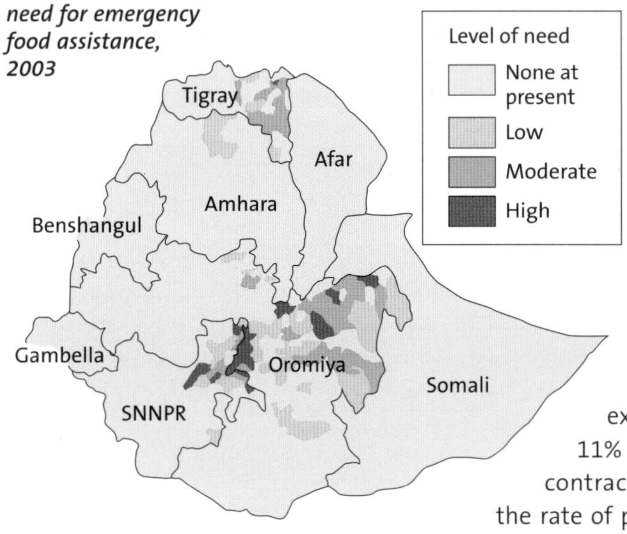

Level of need
- None at present
- Low
- Moderate
- High

The net effect of these and other factors is to preserve and accentuate the **historical pattern** of an unevenly distributed population. As a consequence, carrying capacities in the favoured areas are being pushed to their limits. Indeed, malnutrition and imminent starvation indicate that these limits have already been exceeded in some places (Figure 5). Solutions to this desperate situation are hard to see, except possibly that with an HIV prevalence rate of 11% (one of the highest in Africa) and low use of contraceptives, rising deaths from AIDS might curb the rate of population growth and perhaps even change the distribution pattern.

The points underlined by this case study are that:
- *the physical environment continues to exert a powerful influence on the distribution of population, particularly in an LEDC, because possible means of changing the established pattern are severely limited*
- *non-physical factors also help to perpetuate the pattern inherited from the past*
- *uncontrolled population growth can result in carrying capacities being exceeded, a threshold indicated by the increased incidence of poverty, malnutrition and starvation*
- *there is a need to take into account the vertical dimension in a subject that is generally thought of as being only about the horizontal. Generally speaking, high altitude is thought of as being a negative factor. However, in Ethiopia the situation is rather different.*

Question

(a) **Distinguish between 'distribution' and 'density'.**

(b) **With reference to one LEDC:**
   (i) **outline the main features of its population distribution**
   (ii) **briefly describe the spatial variations in population density**
   (iii) **identify those factors that have had the greatest impact on population densities**

Guidance

(a) Distribution is about the actual location of people within an area. Density is the number of people per unit of land area. A located dot map shows distribution, whereas density is shown on a choropleth map.

(b) Ethiopia is used as the example.
   (i) ■ The distribution is predominantly rural, although 1 in 10 people live in the capital city, Addis Ababa.
   ■ Most of the population lives in the western half of the country.
   ■ 75% of the population live at an altitude between 1500 and 2400 m; 15% live above and 10% live below that range.
   (ii) ■ The mean population density is 55 persons per km². However, at a regional level densities range from less than 30 to more than 100 persons per km².

■ The highest population densities occur in a belt running roughly north–south along the highlands, immediately to the west of the Rift Valley.
■ Much of the eastern part of the country is sparsely populated.

(iii) ■ Altitude is a major influence; population densities are lowest on the highest and lowest ground. Altitude tends to makes its impact through the medium of other factors, particularly climate, carrying capacity and economic opportunities.

Other factors include:
■ disease
■ inertia — the pattern inherited from the past and persistent poverty
■ war — particularly near the Eritrean border
■ urbanisation — particularly rural-urban migration to Addis Ababa

Having read and digested the case study of Ethiopia, it could be tempting to think that the grip of the physical environment on population distribution is going to be much less in MEDCs. After all, their governments have the resources, organisation and technology to intervene and manage. Japan, for example, is one of the most densely populated and economically developed countries in the world. Surely the Japanese have the upper hand when it comes to dealing with the physical environment? The following case study may persuade you to think otherwise.

A WARNING ABOUT DISTRIBUTION

Case study 3

The case of Japan

With a mean population density of 336 persons per km², Japan is one of the world's most densely populated countries. When we look at the distribution at a prefectural level (a prefecture is the Japanese equivalent of a UK county), densities range from 68 persons per km² in the northernmost and rather inhospitable island of Hokkaido to over 6000 persons per km^2 in Tokyo Prefecture, which contains a large part of the capital's metropolitan region.

At face value, the map produced by plotting the density figures for the 47 prefectures (Figure 6) would seem to give an accurate picture of the distribution of population in Japan. However, if you take a look at any physical map of Japan, you will see that it is dominated by mountainous terrain. Although rarely rising above 1500 m, this terrain is deeply dissected by V-shaped valleys separated by narrow ridges. In terms of settlement and development, this upland country has been a persistent handicap, particularly to transport. Even today, it remains largely unpopulated and unused.

In stark contrast to the all-prevailing uplands are the alluvial lowlands, which account for a mere one-eighth of the land area. These lowlands are highly fragmented, small and largely coastal. It is here that most Japanese live and work, mainly in towns and cities. The actual population densities are far higher than the prefectural means shown on Figure 6. So, in a sense, Figure 6 may be accused of concealing reality. Figure 7 gives a more accurate picture and draws our attention to the following key features:

- the relative emptiness of much of inland Japan
- the attraction of population to the coastal lowlands
- the corridor of very high population densities running along the south coast of Honshu from Tokyo westwards and perhaps reaching as far as the northern part of the island of Kyushu. This corridor is occupied by large urban agglomerations that are tending to grow together to form one huge, continuous urban complex (Tokaido megalopolis).

With population densities at levels well over 1000 persons per km^2, it is intriguing to ask why the Japanese have not used their immense wealth and great technological competence, not just to increase the country's carrying capacity, but also to make its population distribution more even. Is it because they recognise that the physical

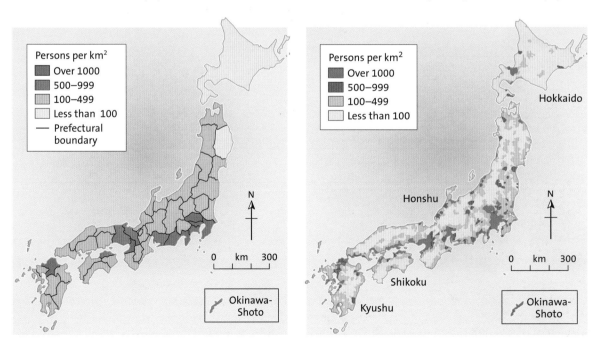

Figure 6 *Japan: population distribution by prefecture, 2000* **Figure 7** *Japan: actual population distribution, 2000*

Contemporary Case Studies

environment is supreme and that in the long run it is better to work with it rather than against it? That may well be the case, but the persistence of the coastal pattern is encouraged by historical factors (the momentum of the past) as well as by economic considerations, such as the importance of the sea to Japan's food supply and economy (importing raw materials and energy and exporting finished goods).

This case study exemplifies a number of points:
- *The physical environment can be an important influence on the distribution of population even in MEDCs.*
- *Strong contrasts in population density can exist within highly populated countries.*
- *The potential deception of choropleth maps as a means of showing the actual distribution of population.*

Change

The shifts implied in the term 'population change' are mainly to do with numbers and structure. A systems view of population sees change as the outcome of two processes — **natural change** and **net migration**. The inputs are births and inward migration (**immigration**) while the outputs are deaths and outward migration (**emigration**) (Figure 8).

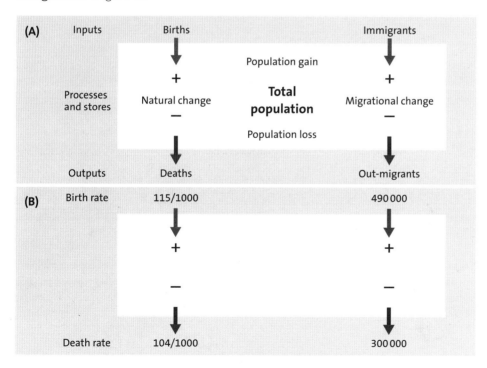

Figure 8
(A) A systems view of population
(B) The UK population system, 2001

- Natural change reflects the balance between births (birth rates, fertility) and deaths (death rates, mortality). Natural increase results when births exceed deaths; natural decrease occurs when deaths exceed births.
- Net migration or migrational change can also be either positive or negative depending on the balance between arrivals and departures.

The nature of the balance between natural change and net migration determines both the direction and rate of population change. Broadly speaking, these two factors can work either together or against each other (Figure 9). If they work together, the level of population can rise or fall significantly. If they are in opposition, one of the components will tend to neutralise the impact of the other. Therefore, the result will be a reduction in the potential scale of population change, be it plus or minus.

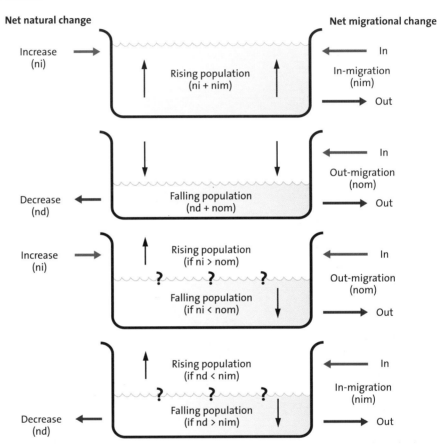

Figure 9 *Different population change scenarios*

Changes in population **structure** are also important. Of the various attributes of structure, age is probably the most significant. The age structure of a population has a direct bearing on birth and death rates and therefore on natural change.

Using case studies 2

Summarising the essential features of a distribution map

The aim here is to build on what you have already done in *Using case studies:1*, namely to improve your ability to describe the essential features of any distribution map. This is a task frequently required in examinations. The global distribution of population change (Figure 10) is a good map to focus on. Of all the global maps showing different aspects of population (density, birth rates, death rates, life expectancy), this is perhaps the most helpful in revealing potential population trouble spots. For example, regions of high population growth can become, if they are not already, regions of overpopulation and overstretched carrying capacities. Conversely, low rates of growth can lead to intolerably high dependency ratios and serious underpopulation. In general terms, the description strategy should be to go for the general rather than the particular. In other words, try to identify the really outstanding features. A useful tip here is to focus first on the 'extreme' areas, namely those showing the highest and lowest values. Deal with each of these in turn by means of a few broad brush-strokes (i.e. bullet points). Pointing out exceptions to the generalisations you have made should be kept to a minimum. Finish by focusing on the middle-value areas.

How does this work out in practice? Let us look at Figure 10 and tackle the question that follows.

Figure 10 *The global distribution of population change, 1990–2000*

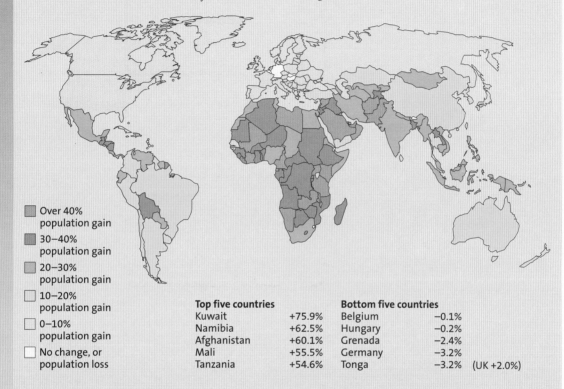

Over 40% population gain

30–40% population gain

20–30% population gain

10–20% population gain

0–10% population gain

No change, or population loss

Top five countries		Bottom five countries	
Kuwait	+75.9%	Belgium	−0.1%
Namibia	+62.5%	Hungary	−0.2%
Afghanistan	+60.1%	Grenada	−2.4%
Mali	+55.5%	Germany	−3.2%
Tanzania	+54.6%	Tonga	−3.2% (UK +2.0%)

Question

Identify the main features of global population change between 1990 and 2000.

Guidance

- High rates of population growth (over 30%) occurred in the South, mainly in an arc that included most of Africa and the Middle East and reached as far as Afghanistan and Pakistan.
- Within that belt, some countries — such as Senegal, Nigeria and South Africa — showed a lower rate of growth.
- Most of the North showed little positive change (less than 10%); indeed, the German and Hungarian populations declined.
- Modest rates of population growth (10–30%) were experienced in India, China, southeast Asia, Australia and Latin America. In the last of these areas, there were a few 'islands' of greater change (for example, Bolivia, Guatemala and Nicaragua).
- Remember that the question has not asked us to explain, just to identify (i.e. describe).

The global map of population change is affected most of all by the balance between birth and death rates. To explain the global distributions of each of these, a whole range of factors needs to be considered (Figure 11).

Figure 11 Some factors affecting fertility and mortality

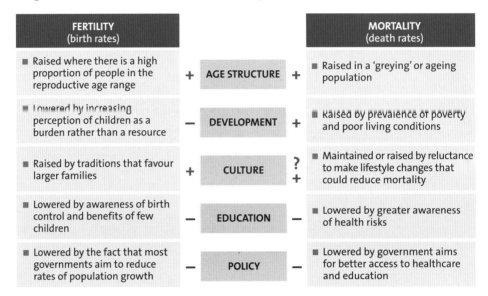

Of these factors, the age structure of a population is particularly important. A diagram much favoured by geographers and demographers — the age–sex pyramid — perhaps best shows the link between age structure and population change (see Figure 12).

A broad-based pyramid tapering fairly abruptly upwards (Stage 2, Figure 12) indicates a youthful population. High rates of fertility and natural increase are to be expected, given the large proportion of people of reproductive age. Conversely, a pyramid shaped like a beehive (Stage 4) indicates an ageing or 'greying' population. Low fertility is reflected in the undercutting of the base, while the gentle taper is the product of long life expectancy and low mortality. In Stage 3, the much narrower base signals a decline in fertility, while the gentle upward taper again suggests that mortality is quite low. A slower rate of population increase is to be expected. Fertility and mortality are so balanced in Stage 1 that a significant change in population numbers is highly unlikely.

The demographic transition model puts these two aspects (changing structure and numbers) into a single time sequence (Figure 12). In the model, countries make the transition from a situation where high mortality in a youthful population produces little growth to one where a similar situation results from low fertility in an ageing population. In between these two states, birth and death rates at first diverge and then converge, producing phases of accelerating and then decelerating growth. These are the critical phases. The model usually recognises four stages, but events in the UK during the last intercensal period (1991–2001) (*Case study 5*) lead us to wonder whether the model might need updating and extending to include a fifth stage (contracting, or underpopulating). On the other hand, recent developments in Russia (*Case study 6*) perhaps suggest the onset of a rather different scenario.

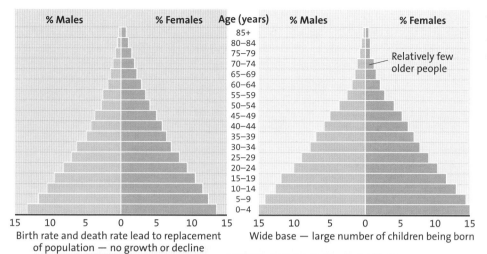

Figure 12
The demographic transition model applied to the UK

Birth rate and death rate lead to replacement of population — no growth or decline

Relatively few older people

Wide base — large number of children being born

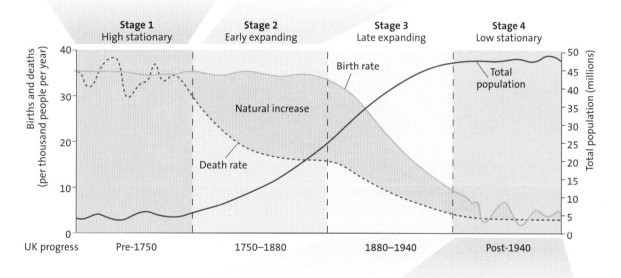

Stage 1 High stationary	Stage 2 Early expanding	Stage 3 Late expanding	Stage 4 Low stationary

Birth rate

Natural increase

Death rate

Total population

UK progress — Pre-1750 — 1750–1880 — 1880–1940 — Post-1940

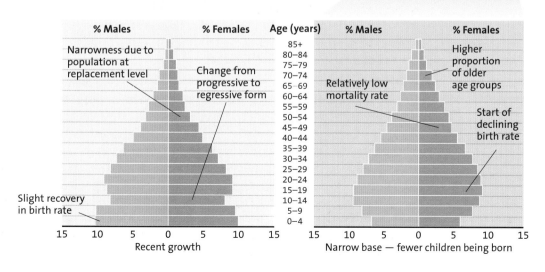

Narrowness due to population at replacement level

Change from progressive to regressive form

Slight recovery in birth rate

Recent growth

Higher proportion of older age groups

Relatively low mortality rate

Start of declining birth rate

Narrow base — fewer children being born

Coping with population pyramids

Population pyramids occur frequently in examination questions on population geography. Candidates are asked to complete any one of at least four different tasks:

■ annotate a pyramid to draw attention to the main features of the population
■ suggest what sort of country is being represented by a particular pyramid — roughly where is it located along the demographic transition model (DTM)?
■ compare the pyramids of two or more populations
■ relate pyramids to different stages in the demographic transition model

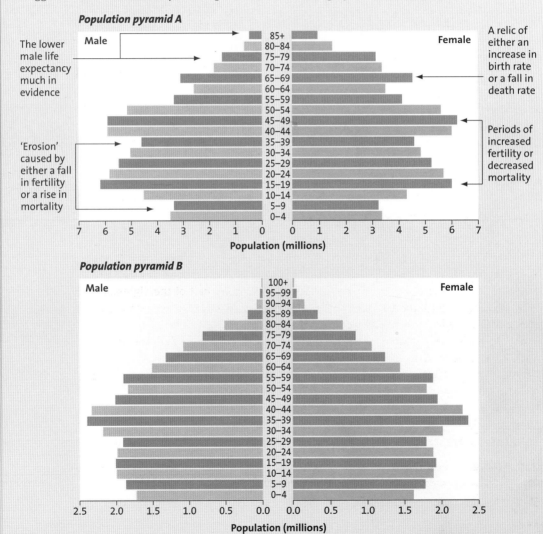

Figure 13
Two population pyramids

Question

Study population pyramid B.
(a) Annotate it to draw attention to its significant features.
(b) What sort of country is represented by this pyramid?
(c) In what ways does pyramid B differ from pyramid A?

Guidance

(a) First of all, how many of you spotted that the horizontal scale used in the construction of both pyramids refers to population numbers? It is more common for each sex age-group to be shown as a percentage of total population. So a tip here — always check whether the sex age-groups are being shown as absolute numbers or as percentages.

Now take a look at the points highlighted on pyramid A. Unless you state otherwise, it is assumed that the observations you make apply to both sides of the pyramid. In this instance, one observation points out an important difference between male and female life expectancy.

(b) An MEDC with a large population that, judging by the 'undercutting' at the base of the pyramid, has reached Stage 4 of the demographic transition model.

(c) Compared with pyramid A:
- the country shown in pyramid B is smaller in population terms
- the population structure has a slightly more regular form
- the undercutting at the base of the pyramid has persisted for longer
- the taper at the top end of the pyramid suggests longer life expectancy (note that the vertical scale in pyramid B extends to 100+)
- the difference in male/female life expectancy is slightly less marked

POPULATION CHANGE AND DEVELOPMENT

Case study 4

Do three Asian examples follow the DTM?

It is often claimed that there is a close relationship between the development process and the progress made by countries through the demographic transition. The UK was perhaps the first country to go through the transition (Figure 12). The driving force was the industrial revolution, which began during the second half of the eighteenth century. By about 1940, the UK had reached the start of Stage 4, having completed Stages 2 and 3 in a little under 200 years (Figure 14A).

The aim of this case study is to examine how three countries in Asia have fared in terms of the demographic transition. Specifically, four questions will be addressed:
- Do these countries appear to have gone through, or to be going through, the same sequence of stages?
- If 'yes', how far have they progressed along the transition?
- How does their speed of progress compare with that of the UK?
- What links, if any, can be found between the DTM and economic development?

The chosen Asian examples are Japan, Taiwan and Thailand. Birth rates and mean annual growth rates are used to compare and monitor their progress along the DTM (Figure 14).

Japan

The industrialisation and urbanisation of Japan began around 1860 (some 100 years after the UK). A little over a 100 years later, Japan had become an economic superpower.

Figure 14 suggests that Japan:
- deviated slightly from the DTM with its steady rate of population growth during Stages 2 and 3 (i.e. between 1880 and 1980); growth appears to have been fuelled by a high birth rate
- progressed through Stages 2 and 3 in about half the time it took the UK

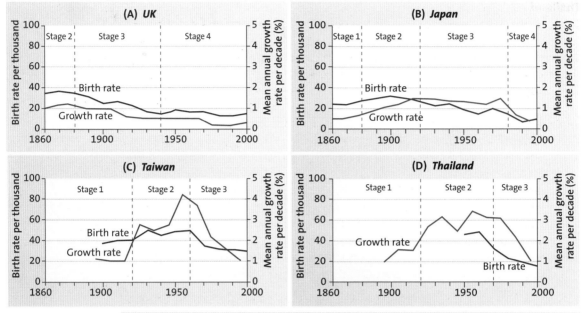

Figure 14
Four demographic transitions

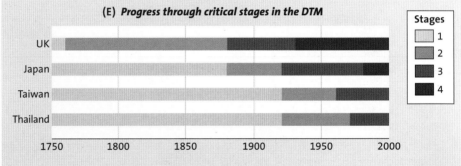

Taiwan

The development process in Taiwan began fairly slowly at the beginning of the twentieth century. At this time, it was a Japanese colony. After the Second World War, thanks to support from the USA, development was based on agriculture and a range of export-oriented light industries. Between 1960 and 1990, the economy grew at an annual average rate of 9.5% and Taiwan joined the ranks of the so-called 'Asian Tigers' — a group of newly industrialising countries (NICs). At present, the country is somewhat in the doldrums. Like others, it is suffering from a 'dose of Asian flu' brought on mainly by the slump in the global economy.

Figure 14 suggests that:
- the critical Stage 2 started around 1920
- a high birth rate persisted to around 1960
- the onset of Stage 3 was marked by the spectacular plunge in the birth rate and in the mean annual rate of population growth between 1960 and the 1990s
- although it is too early to say, Taiwan may currently be entering Stage 4

Thus it seems that not only has Taiwan been running some 40 years behind Japan, it has taken slightly less time to complete the two critical stages of the transition.

Thailand

Thailand belongs to the next generation of developing nations, presently known as the recently industrialising countries (RICs). Between 1920 and 1970, the main feature of economic development was a steady rise in agricultural productivity. During the 1970s, the economy broadened to industries and services. The period from 1985 to 1995 was very prosperous; the average annual growth rate was more than 7.5%. In 1997, the economic bubble burst and this, together with the spread of 'Asian flu', put Thailand into a recession from which it is only just beginning to recover.

Thailand's economic development has not only come later than that of Japan and Taiwan; it also differs in terms of the relative importance of agricultural exports and tourism.

Figure 14 suggests that:

- Thailand possibly started Stage 2 around 1920 (the same time as Taiwan), but the lack of birth-rate data makes this uncertain
- it was not until around 1960 that the growth rate began to fall from a high level of 3% per annum, thus marking the graduation from Stage 2 to Stage 3
- although there are clear signs that the rate of population growth is continuing to fall, it cannot be said for sure that the country has yet reached Stage 4

Thailand appears to have moved slightly more slowly along the demographic transition than Taiwan. In both instances, the graduation from Stage 2 to Stage 3 seems to have coincided with a significant take-off in economic growth — during the 1960s in Taiwan and during the 1970s in Thailand. Equally, the move towards Stage 4, as evidenced by the flattening out of population growth rates, looks to have been hastened by the global economic recession of the late 1990s that hit Asian countries so badly.

4

Using case studies

Here is a task for you to try yourself.

Question

Read the above case study, paying particular attention to Figure 14 (especially E).
- (a) **Do all four countries appear to have gone through, or to be going through, the same broad sequence of demographic change?**
- (b) **How do their rates of progress through the model compare?**
- (c) **How might you explain any differences in the rate of progress?**
- (d) **What demographic change appears to mark the beginning of:**
  - (i) **Stage 2?**
  - (ii) **Stage 3?**
- (e) **Explain the possible links between the demographic transition and economic development.**

THE DAWN OF STAGE 5 IN THE DTM?

Case study **5**

The UK today

Between 1870 and the beginning of the Second World War in 1939, the UK's birth rate fell from 37 to 12 births per 1000 population, so taking the country into Stage 4 of the demographic transition (Figure 12). That fall can be explained in terms of:
- the increased use of birth control methods
- more employment opportunities for women

- the rise of an urban society that no longer saw any advantage in having large numbers of children
- the growing perception of children as a drain on household finances and an impediment to socio-economic progress

Figure 14A shows that throughout the second half of the twentieth century, the birth rate fluctuated between 13 and 18 per 1000. At the last census (2001), it stood at 12 per 1000. Is the persistence of low fertility in the UK still explained by the same four factors?

The short answer is that those explanatory factors remain valid and may even have been strengthened. For example, more reliable and readily available contraception means more effective birth control. British society has become even more urbanised, and the costs of housing and child-raising have rocketed. For increasing numbers of people, a career comes before children. However, as Figure 15 shows, these are not the only factors involved.

Figure 15
Factors reinforcing the persistence of low birth rates

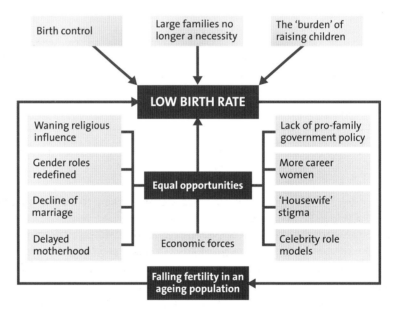

The progressive 'ageing' of the UK's population is reducing the percentage of the population that is of reproductive age, so falling birth rates set in motion a downward spiral of even lower birth rates (Figure 15). The second additional factor is a set of changes in British society linked, to varying degrees, with legislation put in place over the last 50 years to end sex discrimination and enforce equal opportunities. These changes are outlined below:

- **gender roles redefined** — exemplified by the advent of working women and 'house husbands'
- **declining incidence of marriage** — the increasing popularity of couples living together outside marriage and without children
- **delayed motherhood** — many women who choose to have children are postponing their first pregnancy until they are in their thirties and have had time to forge a career; this often results in one-child families
- **economic pressures** — the high costs of housing, living and raising children

- **government policy** — little pro-family emphasis
- **more career women** — increasing numbers of women are foregoing both marriage and having children
- **the 'full-time housewife' stigma** — women complain that their peers look down on them if they give up their careers and stay at home to raise children
- **celebrity role models** — tend to promote the 'glamour' of having a career and of returning to work immediately after giving birth, making use of child-minding services

In the second part of *Case study 1*, reference was made to a recent increase in the UK's population. Currently, the growth is explained by rising immigration rather than by natural increase. However, given the nature of many of the new immigrants — essentially young people drawn from societies where traditional attitudes about large families persist — it seems likely that Britain's birth rate may soon be lifted out of the doldrums.

The steady rise in life expectancy (looked at more closely in *Case study 22*) continues for most sex/age groups in the UK.

Much of the reduction in British mortality rates is due to improvement in the survival rates among people afflicted by the four main killer diseases. During the 1990s:
- deaths from coronary heart disease fell by 36% among men and 40% among women
- lung cancer deaths fell by 28%
- the incidence of breast cancer in women fell by 24%
- the number of stroke victims fell by 30%

A government prone to spin would claim that this is the outcome of a 'massive' increase in spending on the NHS. In 1990, the UK was spending 6.2% of GDP on healthcare; in 2000 the figure was only 7.1%. In contrast, France spent about 9.3% per annum over the same period. The lower mortality figures are more likely to be the outcome of education, changing lifestyles and earlier diagnosis.

With fewer births and deaths, the overall population becomes still 'greyer'. This is reflected in the fact that 15% of the population is now aged 65 and over. The slight decline in the dependency ratio from 0.62 in 1971 to 0.55 in 2001 conceals the fundamental shift in the character of dependency (Figure 16). The percentage of children in the population has slipped to 20%. Projections suggest that by 2014 there will be more over-65s than under-16s.

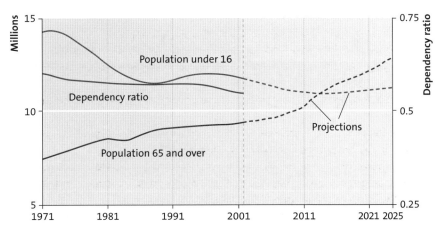

Figure 16 Changing dependency in the UK, 1971–2025

If we believe that the DTM is in the process of developing a fifth stage, then the following demographic characteristics, shown in this case study, could be among the hallmarks:

- *the lowering of an already low birth rate*
- *first-time mothers becoming older*
- *the rise of childless marriages and partnerships*
- *a change in traditional lifestyles, such as more same-sex relationships*
- *a rise in the death rate, despite health improvements as the population increases in 'greyness'*
- *changing dependency*
- *the potential decline in population offset by immigration*
- *immigration increasing ethnicity*

One question arising from this case study is: do the explanations of persistently low birth rates and declining death rates in the UK apply to other countries? For example, how is it that Catholic countries such as Italy and Spain (Case study 8) now have the lowest birth rates in Europe? How is it that the UK mortality rate has been falling more quickly than that of other western European countries?

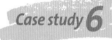

Case study 6 — POPULATION GOES INTO FREEFALL

The Russian experience

Russia is facing a demographic crisis as the live birth rate plummets and the death rate is raised through excessive consumption of tobacco and alcohol. Today, less than one-third of recorded pregnancies produce live births. The plunge in the live birth rate has been brought about largely by illnesses (e.g. anaemia, heart problems and malnutrition) among pregnant women and by abortion being used as a form of birth control.

Figure 17
Poverty on the streets of Moscow, contributing to Russia's demographic downward spiral

Topfoto

Adding to the agony is the high rate of infant mortality. Recently released figures indicate that 30% of all Russian newborn babies die of infections. A large number of babies born with correctable heart defects are not being operated on. Healthcare has been one of the worst victims of Russia's transition from a socialist to a market economy. The system has crumbled. At the same time, cigarettes and vodka have lowered life expectancy to among the shortest in the world (Figure 17). Average male life expectancy is now only 59 years — 10 years less than the mean expectation in western Europe.

As a consequence of the decline in live births and the increases in infant and adult mortality rates, the total population of Russia fell by 6 million during the 1990s from a peak of 148 million. It could fall by another 39 million by 2025.

The immediate question raised by this case study is whether or not the Russian predicament is a one-off situation. If not, is Russia to be seen as entering Stage 5 in the DTM, but one that is perhaps rather different from that facing the UK? Or does the Russian experience simply represent a regression back to an earlier stage in the model?

Using case studies 5

Look back at the population pyramids in Figure 13. Population pyramid A shows the Russian Republic and pyramid B shows the UK.

Question

The UK and the Russian Republic may be two of a small number of countries pioneering what is now being seen as Stage 5 in the DTM. Using *Case studies 1–6*, compare their present demographic characteristics. What conclusions do you reach?

Guidance

- The pyramids in Figure 13 suggest that the population trends have been more consistent in the UK; volatile change is evident in the Russian pyramid.
- The UK population is rising slowly; that of Russia is falling.
- The increase in the UK can be explained by immigration; possibly emigration is playing its part in Russian decline.
- Birth rates are declining in both countries.
- The death rate is falling slowly in the UK, but rising in Russia; the rise in infant mortality is particularly marked.
- There are some major differences here and it is difficult to believe that both countries are passing through the same demographic transition phase. The evidence provided by Russia suggests that it may be possible for countries to regress to an earlier stage.

Part 2

Migration and consumption

Migration

In studies of population, the term movement (or **mobility**) has two components — migration and circulation. Both involve people shifting locations. **Migration** refers to those moves involving a change of residence that lasts for at least 1 year. **Circulation** involves moves that are shorter term; the changes in location are in a sense temporary. Shopping, commuting, tourism, pastoral nomadism and shifting cultivation are examples.

It will only take you a little time to come up with a list of population movements. No doubt your examples will be drawn from different parts of the world. It is also likely that they will differ in terms of scale (i.e. numbers of people and distances

Figure 18
A classification of population movements

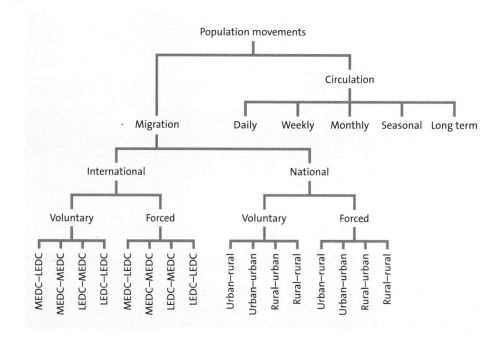

Contemporary Case Studies

involved), when they occurred and for what reasons. Figure 18 suggests one possible classification that helps us to put these and many more examples into a single framework. Migration and circulation represent the first-order subdivision. Circulation is then subdivided on the basis of its frequency. Equally, it could be broken down on the basis of its purpose, as suggested by the examples in the previous paragraph. The classification of migration is rather more complicated. There are many possible criteria:

- distance
- direction
- volume
- cause
- motivation
- the nature of the decision-making process

The distinction between national and international migration is important because it acknowledges the impact that state boundaries can have on population movements. At the next level, the distinction between forced and voluntary migration is thought to be of great significance.The final tier in the classification emphasises the significance of direction, for instance between urban and rural areas within a country and between LEDCs and MEDCs at a global or international level.

Figure 19
The main global migrations of the last 25 years

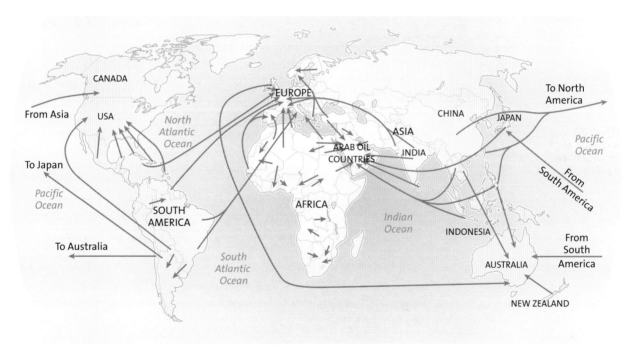

Figure 19 shows the main global migration flows of the last 25 years. Geographical studies of these and lesser migrations seem to have two main focuses: motives and impacts. There are at least two sides to each of these focuses. In the case of motives, there are both push and pull factors to be considered. The impacts of migration on the **source region** are very different from those on the **destination** or **host region**.

There is also a third side to impacts, namely the effects of the move on the migrant. What sort of reception does the migrant receive? Are migrants welcomed and how easy is it for them to become assimilated into the host society?

It is possible to produce a conceptual framework for migration studies by a cross application of the motives and impacts subdivisions. This yields a matrix of six compartments (Table 1). The case studies that follow illustrate each of these six different contexts.

Table 1
Case studies of six different migration scenarios

Motive	Impacts on source	Impacts on destination (host)	Impacts on migrants
Push	(7) South Africa	(9) Trinidad	(18) The Kurds (28) The UK and asylum seekers
Pull	(7) South Africa	(8) Rural Spain	(10) UK immigrants

6 Using case studies

Question

Summarise the main global migrations of the last 25 years, as shown in Figure 19.

Guidance

- 'Summarise' is the vital command word. It means identify the main flows. Do *not* indulge in a tedious flow-by-flow description. The mark allocation, together with the lined space allocation, should indicate the degree of generalisation required by the examiner.
- Focus first on the main destinations of intercontinental migration, as indicated by the convergence of the flow lines. These are the USA, Europe, the Middle East and Australia.
- If there is space, outline the main source regions for each of these destinations.
- Africa deserves a quick mention in that there was a fair amount of intra-continental migration; to a lesser extent, so does South America.
- Remember you have only been asked to summarise, *not* to explain, what you see on the map.

Reasons for migration

It is no exaggeration to say that migration is one of today's major global issues. The growing volume of international migration draws our attention to the persistence of unbearable problems in the main source regions. Equally, these same flows are creating some serious problems in the more popular destination regions. It is widely held that most migration is the outcome of two sets of forces (Figure 20).

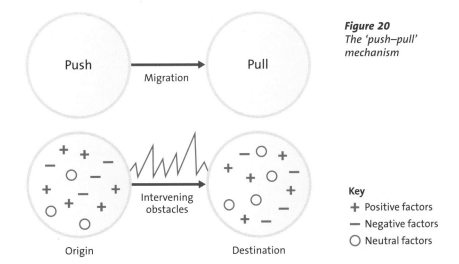

Figure 20
The 'push–pull' mechanism

Key
+ Positive factors
− Negative factors
○ Neutral factors

Push forces work in the migrant's current location. They can range from unemployment to persecution; from natural disasters and famine to poverty and war. The **pull forces** are those that attract the migrant to a particular location. They are mainly the mirror images of push factors and therefore include attractions such as job opportunities, tolerance, personal safety, freedom, good housing and welfare services. Intervening obstacles, both perceived and real, need to be overcome before migration takes place. They include international frontiers and the often considerable costs of moving. Facilitating factors that encourage migration include culture, language, the media, perception and ease of access to visas and permits.

The decision to move will rarely be made on the basis of a single push or pull factor. Rather, the decision will be based on an appraisal of a range of attributes, both in the place of origin and at the potential destination. Each person perceives these attributes differently, depending on personal characteristics such as age, sex, socio-economic class, occupation and education. Some of the attributes at the present location will be regarded positively, persuading the person to stay put. Some will be seen negatively, encouraging migration. Other attributes will be perceived neutrally and thus have no bearing on the decision making. The same threefold classification applies to the potential destination area, except that here the positive factors encourage, and negative factors discourage, migration.

7

Using case studies

Here is another exercise based on Figure 19.

Question
The flows shown on Figure 19 have not been classified.
(a) Attempt to distinguish between those migrations that were essentially voluntary and those that were forced.
(b) What appears to have been the most powerful push factor?

Over the last 150 years, the concept of push and pull has been refined, and a range of other ideas has been applied to the study of migration. They are summarised in Table 2.

Table 2
A chronology of different ideas applied to the study of migration

Name	Date	Key ideas
Ravenstein's laws of migration	1875–89	Most migrants travel short distances and with increasing distance the numbers of migrants decrease Migration occurs in a series of waves or steps Each significant migration stream produces, to a degree, a counterstream Urban dwellers are less migratory than rural dwellers The major causes of migration are economic
Stouffer's theory of intervening opportunities	1940	The volume of migration between two places is related not so much to distance and population size, but to perceived opportunities that exist in those two places and between them
Zipf's inverse distance law	1949	The volume of migration is inversely proportional to the distance travelled by migrants, and directly proportional to the populations of the source and destination
The gravity model	1960s	This simple formula expresses Zipf's two relationships
The Lee model	1966	This revised the simple 'push–pull' model in two ways It introduced the idea of 'intervening obstacles' that need to be overcome before migration takes place Source and destination are seen as possessing a range of attributes; each would-be migrant perceives these attributes differently, depending on personal characteristics, such as age, sex and marital status
The Todaro model	1971	This stresses that potential migrants weigh up both the costs and benefits of moving before taking any action; migrants act in economic self-interest
The Stark model	1989	This extends the Todaro model by arguing that there is more to migration than the optimising behaviour of migrants; risk spreading in families is one such factor
Marxist theory	Fashionable in the 1980s	Migration is seen as the inevitable outcome of the spread of capitalism Migration is the only option for people once they are alienated from the land
Gender studies	1990s	These emphasise that men and women differ in their responses to migration factors and that sex discrimination in the labour market has an important impact

Case study 7 EMIGRATION FROM SOUTH AFRICA

Reasons and consequences

Emigration has emerged as one of the greatest challenges to South Africa since the end of apartheid in the early 1990s. Skilled South Africans are finding it increasingly attractive to work abroad, where their skills are eagerly snapped up. The outflow is seriously threatening the country's efforts to raise its rate of economic growth.

It is estimated that 30 000 South Africans left the country in 2001 to join the 1.6 million already living and working abroad. The chances of more leaving are high. As in most developing countries, emigration from South Africa is essentially a brain drain, or exodus of skills. More than half of emigrants are professionals, semi-professionals, managers or executives. Typically they include doctors, lawyers, accountants and engineers, but their numbers are swelled nowadays by young people with IT skills, teachers, nurses, farmers and people with particular manual skills. It is estimated that more than 20% of South Africa's professionals have already left. Some 70% of skilled South Africans are currently considering emigrating. The causes of this exodus are many, with the main ones set out in Table 3. Some of these 'push' factors have mirror images that serve to reinforce the wish to migrate.

Reason	Explanations
Violent crime	Reason cited by 60% of emigrants
	South Africa now ranks alongside Russia and Columbia as one of the most dangerous countries in the world
	During the 1990s, a total of 250 000 people were murdered; on average, 750 000 violent crimes were reported annually
The economy	Reason cited by 10% of emigrants
	Particular factors include: ■ the huge devaluation of the South African rand ■ the high rate of income tax ■ high interest and inflation rates ■ increasing difficulty of many white people to find the kind of work that matches their skills
	The next three uncertainties tabulated below are also discouraging investors
Mbeki's two-nation approach	The shift away from Nelson Mandela's reconciliatory approach to whites to an emphasis on race under President Mbeki
	Mbeki's belief is that South Africa consists of a rich white nation and a poor black nation and that economic sacrifices will have to be made by white people
	Many white people are nervous as to what those sacrifices might involve
AIDS	South Africa now has one of the highest rates of HIV infection in the world
	The strain this is putting on healthcare facilities, the debilitating impact of the disease on the workplace and the fear of contracting it are aspects that help to push the emigrants
Zimbabwe	The crisis in neighbouring Zimbabwe has unnerved many white South Africans, who fear that the situation may spill over into their country
The global village	In an era of globalisation, many emigrants simply wish to transfer their skills overseas to earn higher incomes
	Some of these economic migrants eventually return, but for others the global village becomes their new home

Table 3
South Africa: reasons for emigrating

Destination	% of emigrants
UK	25
Australia	18
USA	11
New Zealand	10
Canada	7
Western Europe (excluding UK)	25
Others (e.g. Switzerland, Israel, Namibia)	4

Table 4 *South Africa: destinations of emigrants*

Just over 70% of emigrants move to five countries, all of them English-speaking MEDCs with cultural similarities to, or historic ties with, South Africa (Table 4). The UK is the most popular destination because around 800 000 South Africans have, or can lay claim to, British passports. Britain allows visa-free visits for South Africans and offers 2-year working visas for South Africans who are under 27 years of age. There is no question that the influx of South Africans into the UK is helping fill gaps in the skills market, for example within the National Health Service.

This brain-drain costs South Africa about £159 million a year, with the loss of each skilled professional resulting in the loss of as many as 10 unskilled jobs. The loss — particularly among young people — is felt most acutely in engineering, medicine, accounting and financial services. A shortage of skilled labour is now thought to be one of the greatest threats to South Africa's future. Clearly, South Africa's loss is a gain for the destination countries.

'South Africa is shedding skills at a worrying rate to its global competitors. These countries have no compunction about creaming off skilled people from other countries,' says one South African official. So how is the South African government to react? Prohibit the emigration of mainly white workers? Train local black people to fill the gap left by emigrating professionals? Encourage the immigration of foreigners? Another official has urged that 'if South Africa is to enjoy widespread long-term economic growth, it has to open its doors to foreign skills.'

Points raised by this case study include:
- *The desire or need to emigrate is not confined to cheap, unskilled labour in search of work.*
- *A nation is only likely to lose its best workers when powerful push factors prevail.*
- *The movement of economic migrants can have serious cost implications for the source country, but considerable benefits for the receiving country.*
- *Should governments intervene to stop such costly movements?*
- *Is globalisation really a beneficial process when it allows labour to be drawn away from one country, only to be replaced by workers from another?*

Two important points about the reasons for migrating need to be fully understood before moving on to consider the impacts of migration:
- The decision whether or not to migrate is largely conditioned by a number of factors that are personal to would-be migrants. These include their circumstances, their perceptions of opportunity and risk, and their values.
- In analysing the reasons behind the decision to migrate, it is easy to make assumptions that misinterpret motives. For example, the tabloid press in the UK prefers to see the current influx of immigrants from Eastern Europe as being made up largely of people intent on 'sponging on the state'. In reality, it is much more likely that those people are driven here by poverty, the need to find work (they have skills that are in demand) and the desire to start a new iife. In short, migration is a topic open to bias and we need to be alert to this danger.

Much of what we read, whether it be a textbook or magazine article, has an in-built slight bias that reflects the values and perceptions of the writer. However, there are contentious aspects of population and migration in which there is scope for bias of a more sinister kind. This seeks to persuade the reader to see things in a particular way, which may be removed from both reality and the truth.

Question

Read the following extract taken from the editorial column of the *Daily Express* for Thursday 26 February 2004:

The mass exodus of Britons to Spain exposed by the *Daily Express* today is the latest sign of growing disillusionment with British life in the twenty-first century.

Illegal immigrants and asylum seekers view these shores as a land of milk and honey, but for Britons who have given up on the charms of this country, there appear to be no regrets. Our report into the life enjoyed by expatriates shows that Spain offers more than just sun and sea. It also offers superior standards in terms of salaries, safe streets, better schools and state-of-the-art hospitals.

Even some of the immigrants coming here are only using British citizenship as a stepping stone to somewhere better.

While bringing up the standard of living in all new EU countries is something to be encouraged, not least because it will reduce the incentives for migrants to head to Britain, the government cannot just sit back and allow the exodus of some of our most skilled workers to continue unchecked.

Research the factual basis of this extract and expose its bias.

Guidance

It may be helpful to use some of the information in Figure 26.

Impacts of migration

Just as each population movement has its causes, so each migration and circulation has its consequences or impacts. These fall into a number of categories.

Demographic impacts

The most obvious outcome of migration is redistribution of population. One location's gain is another's loss, and vice versa. Other consequences are related to the fact that migration is selective. For example, because it tends to involve people of reproductive age, migration impacts on birth rates. For receiving or host locations, this effectively means a double whammy — newcomers, plus a raised rate of natural increase. Conversely, for source locations, the loss of people is reinforced by a dip in birth rate and a decline in natural increase.

Economic impacts

Much migration appears to be triggered by economic considerations such as work opportunities, wage and salary levels. What this means is that in-migration helps to boost locations that are already on the way up in terms of prosperity. That prosperity represents an irresistible attraction for many migrants, particularly those living in poorer and more deprived parts of the world. Out-migration can start, and aggravate, downward spirals of decline in source areas. Equally, the trickle back of remittances from emigrants can be a major source of income for family and relatives left behind.

The next case study focuses on an unusual turnaround in migration and sets out the hoped-for demographic and economic consequences.

Case study 8

REPOPULATING RURAL SPAIN

A new life in the Old World

The migration history of Spain shows two major U-turns. For centuries, up to the last quarter of the twentieth century, Spanish migration was about emigration. Generations of Spaniards struggling to make a living perceived the Americas as the 'promised land'. However, after the Second World War the migration pattern changed completely, with three-quarters of emigrants heading for the countries of northern Europe. Up to 100 000 Spaniards were leaving every year.

An even more profound shift in the pattern of migration occurred after 1970, when immigration became the dominant flow. In 2000, for example, 90 000 new immigrants arrived, while emigration dried up to a trickle of 2000 people. The reason for the U-turn was the take-off of the Spanish economy. Having exported its labour force for so long, Spain suddenly found itself desperately short of workers. A low and dropping birth rate meant that natural increase would not meet the shortfall. The only solution was to import cheap labour, mainly from Morocco and other parts of Africa that had once been Spanish colonies. Those described as 'foreign residents' now account for nearly 3% of Spain's population.

While Spain has had no option but to let in foreign workers, it has struggled to keep immigration under control. Its wish has been to regulate the in-flow to ensure that immigrant skills closely match those required by the economy. In the event, the Spanish authorities have become concerned about the nature of the immigrants. In their view, too many have been coming from North Africa and they are often ill suited to the specific needs of the service sector vacancies. The authorities have also become alarmed by the rising tide of illegal immigrants. The Spanish have discovered what many other countries have before: the more you tighten up on immigration, the more this seems to encourage illegal entry.

A dramatic turn of events occurred in January 2003 when the Spanish government passed a law encouraging various categories of 'non-citizen' to apply for passports. These 'non-citizens' are mainly the descendants of two groups of Spanish emigrants: those who settled in Latin America and those who moved to other European countries as refugees to escape the fascist regime of General Franco (a dictator who ruled Spain from 1936 to shortly before his death in 1975). Both groups were suddenly deemed to be Spanish. The upshot is that around 1 million people may have the right to return to Spain — 850 000 from Latin America and 150 000 from Europe. Spanish embassies in Latin American countries have been inundated by tens of thousands of applicants for Spanish passports. The biggest queues have been in Argentina, a country torn apart by economic crisis and home to the largest overseas Spanish community.

This sudden U-turn has given momentum to an initiative taken a few years earlier by some local authorities in Spain to halt rural depopulation. Since the 1950s, about 2000 remote settlements across Spain have been abandoned as a result of emigration and rural–urban migration. However, the tide may be beginning to turn in the remote mountainous area of Teruel Province to the east of Madrid. This is the least populated part of Spain, one of the poorest and the only one where deaths exceed births. For several years now, village authorities have been offering Latin Americans free plane tickets, jobs and houses to encourage them to move to this isolated area, in a last desperate attempt to stop their settlements disappearing altogether. Settlers are also being sought in eastern Europe.

The leader in this search for new blood has been the village of Aguaviva (Figure 21). In 2000, its mayor set up the Association of Spanish Villages against Depopulation. Membership has already risen to 130 communities. The Association is looking for married couples under 40, without college degrees, with work permits (or one of the new passports) and at least two children under 12. Qualifying families sign a 5-year contract. In return, they are given jobs paying between £425 and £600 a month, housing at a monthly rent of £100, free healthcare and

Figure 21 *The village of Aguaviva, Spain — leading the hunt for new settlers*

schooling. The available work is mainly in farming and services, but it is recognised that any lasting revival of the villages will need the creation of more jobs, particularly for women, the young and people over 45. Of course, it is hoped that after the 5 years are up, families will be happy to stay on.

Most of the recruiting for new settlers has taken place in Argentina and Uruguay, but Aguaviva now has 100 new residents from Romania. In this village, the effects of the scheme are already visible. The average age of the local people sitting in the village square's bar taking an afternoon drink is 82, while that of the incomers is 31.

Thus, centuries after Spain sent its sons and daughters to colonise the New World (the conquistadores), it is now trying to persuade their descendants to return and help to revive Spain's dying country villages. A truly curious turn of fortune!

This case study illustrates the 'swings and roundabouts' that mark one country's migration history. Change in the balance of push and pull factors has led to a complete reversal of the dominant migration direction. While migrants themselves are decision makers, the journeys they make are conditioned by the decisions and policies of governments, both national and local.

Question

Read *Case study 8* again and produce an annotated time-line diagram that summarises Spain's main migration trends since c.1900.

Guidance

Figure 22 might give you some ideas.

Social impacts

The prospect of a better life is probably the single reason most mentioned by migrants when asked why they have moved. Happily, for most migrants the dream of a better quality of life is realised. However, there will also be social costs, such as the stress involved in adjusting to an unfamiliar and sometimes hostile environment. Tensions between 'natives' and 'newcomers' are commonplace, while initially the latter often form an underprivileged class obliged to do menial jobs and to occupy poor housing.

Cultural and ethnic impacts

Throughout history, migration has played a key role in the spread of different cultures (particularly language and religion) around the world. Few would doubt that this spread has been enriching. However, conflict often arises when people of a particular cultural or ethnic background move to a location where they form an 'alien' minority. This is particularly so where intolerance prevails and where the immigrant and the host community feel in some way threatened. The need for mutual security explains why immigrants of the same ethnic or cultural background are often found living in the same area.

Environmental impacts

Migration can give rise to adverse environmental impacts at both ends of the movement path. At the departure end, it is the processes of abandonment, decline and neglect that scar the environment, whereas at the arrival end, the key processes are more likely to be congestion and the overstretching of resources.

Political impacts

The political impacts of migration can range from the growth of empires (like those of European nations during the seventeenth, eighteenth and nineteenth centuries) to the global spread of terrorist organisations such as al-Qaeda. The movement of people within a country often leads to governments having to face the complex issue of helping both core and peripheral regions. Such internal movements can also alter voting patterns. A final point here is not so much about impacts as about the role that governments can, and do, play in both encouraging and discouraging migration. Such political interference is made both possible and necessary by the fact that much migration involves crossing national frontiers. Tied up with all this are the complex issues raised by asylum seekers and refugees (see *Case study 28*).

It is important to note that, under each of the bullet points about the impacts of migration, there are normally both 'pluses' (benefits) and 'minuses' (costs) and that these costs and benefits often co-exist at both the source and the destination.

The next case study is partly about reasons, but also illustrates a number of the cumulative impacts of migration.

PUSH, PULL AND IMPACTS

Trinidad: a multi-ethnic society

Trinidad, a West Indian island just off the coast of Venezuela, has a population of 1.2 million made up of at least five officially recognised ethnic groups — African, Indian, Chinese, white and 'mixed'. This multi-ethnicity is the outcome of a long and varied migration history (see Figure 22).

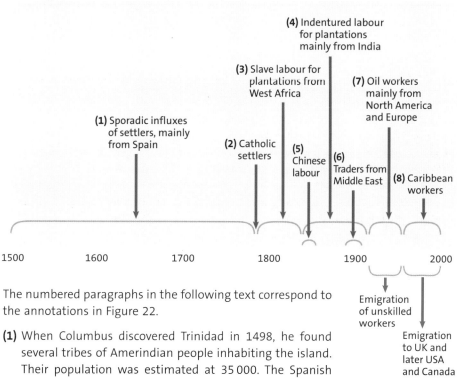

Figure 22 *Trinidad: a chronology of immigration*

The numbered paragraphs in the following text correspond to the annotations in Figure 22.

(1) When Columbus discovered Trinidad in 1498, he found several tribes of Amerindian people inhabiting the island. Their population was estimated at 35 000. The Spanish settlers, who followed Columbus, brought with them diseases (especially smallpox) and were harsh in their treatment of the Amerindians who, consequently, largely died out. However, even to this day, a small number of people in Trinidad claim to be descendants of the native people.

(2) The real development of Trinidad did not begin until the end of the eighteenth century when the King of Spain offered free land to citizens of any country friendly to Spain, provided they were Roman Catholic. This meant that most of the new settlers came from France, since England, Spain's other ally at the time, was mainly Protestant.

(3) The 'pull' of free land was strengthened by the opportunity it gave to the settlers to establish plantations growing a range of crops for the European market — mainly sugar, but also coffee, cocoa, cotton and citrus fruits. However, the plantations needed cheap labour. Before long, people were being rounded up in western Africa and forcibly transported (i.e. 'pushed') to Trinidad as slaves. Between 1783 and 1789, at least 10 000 arrived on the island. This immigration and the spread of plantations had a major impact on the environment as forests were cleared for cultivation and settlements. In 1797, Britain acquired the island as one of its colonies. (Trinidad remained a colony until gaining its independence in 1962.) Under British rule, the plantations expanded and all seemed well until 1834 when the British Parliament passed an act abolishing slavery. This caused much consternation among the plantation owners, but they eventually agreed to free all the slaves on the island. Some slaves left, but most stayed.

(4) As a consequence, the plantation owners were forced to find an alternative source of cheap labour. Eventually they turned to India. Under the indenture scheme, labourers were recruited for 5 or 7 years. At the end of this time, they could choose to be either repatriated or given some land and allowed to stay. By the time the Indian government banned emigration to Trinidad in 1917, the number of indentured workers had risen to more than 145 000. The significant difference was that immigration was now voluntary rather than forced. Many preferred to stay on. Today, 'East Indians', as they are called, constitute around 40% of the total population, about the same as those of African extraction.

(5) Another group that found its way to Trinidad was the Chinese. They were brought in when there was a temporary halt in Indian immigration. The Chinese were not a great success as plantation workers. Soon they moved off to open small businesses of their own, forming a close-knit little community of active entrepreneurs. Today, their descendants account for around 1% of the population.

(6) Also involved in the commercial life of Trinidad today are people of Lebanese and Syrian origin. Their ancestors arrived mainly during the nineteenth and early twentieth centuries.

(7) The next phase in the migration story coincides with the first exploitation of oil (initially from the famous Pitch Lake) in the early part of the twentieth century. By the mid-1920s, oil was the number one export. The expansion of the oil industry drew in more labour: technicians, mainly from North America and Europe, and unskilled labour from local sources and other West Indian islands.

(8) Today, the level of immigration is low. Probably up to 75% come from other Caribbean islands, apparently 'pushed' by high rates of unemployment and 'pulled' by the chance of work in service industries.

This long and varied immigration saga clearly explains the rich ethnic mix that characterises Trinidad society today. To their credit, the groups generally live in harmony. In fact, some 15% of today's population is classified as being of 'mixed' origins. However, when it comes to looking at the impacts of immigration, some differences between the ethnic groups begin to emerge (Figure 23).

The migration story of Trinidad is not just one of immigration. During the twentieth century, there were two significant exoduses of mainly unskilled workers: before and then after the Second World War (1939–45). For example, during the period 1980–90, 52 000 people emigrated, 58% of them to the USA and 25% to Canada. In view of the historic links with the UK, it is surprising that less than 10% of these emigrants headed this way. Thirty years earlier, the situation was rather different. However, as British economic interests have turned away from its former colonies, so the influence of the USA and Canada has grown.

No matter the destination, most emigration has been 'pushed' by periodic slumps in the Trinidadian economy (particularly in the oil industry) and 'pulled' by employment opportunities and the perception that a 'better life' is to be found outside the island. Most emigrants are likely to be unskilled and obtain only low-paid jobs. The main concentration seems to be in hotels and in public services such as hospitals. Illegal immigrants often find that they are vulnerable to excessive exploitation by ruthless employers.

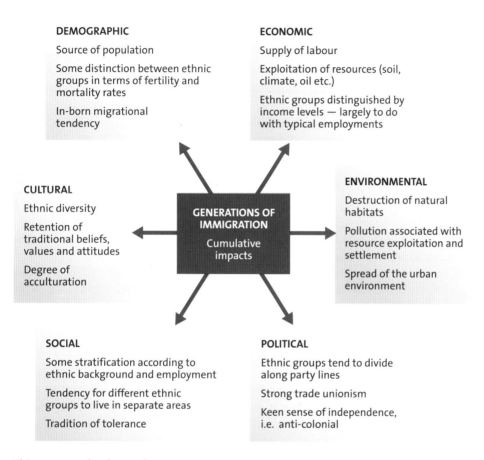

Figure 23 *Trinidad: some impacts of immigration*

DEMOGRAPHIC

Source of population

Some distinction between ethnic groups in terms of fertility and mortality rates

In-born migrational tendency

ECONOMIC

Supply of labour

Exploitation of resources (soil, climate, oil etc.)

Ethnic groups distinguished by income levels — largely to do with typical employments

CULTURAL

Ethnic diversity

Retention of traditional beliefs, values and attitudes

Degree of acculturation

GENERATIONS OF IMMIGRATION

Cumulative impacts

ENVIRONMENTAL

Destruction of natural habitats

Pollution associated with resource exploitation and settlement

Spread of the urban environment

SOCIAL

Some stratification according to ethnic background and employment

Tendency for different ethnic groups to live in separate areas

Tradition of tolerance

POLITICAL

Ethnic groups tend to divide along party lines

Strong trade unionism

Keen sense of independence, i.e. anti-colonial

This case study shows that:

- *the modern population of Trinidad is almost totally migrant in its origins. Much of the explanation of how and why Trinidad is as it is today lies in a long and varied migration history.*
- *over the centuries, economic forces have been paramount in explaining the movement of people into and out of the country. In this respect, the Trinidad story is in no way exceptional.*
- *the ethnic diversity of Trinidad's immigrants is distinctive*
- *despite years of integration and living in reasonable harmony, the ethnic groups tend to remain distinct. Each tends to play a particular role in modern society, retaining much of its traditional culture. This is the most profound and enriching impact that immigration has had on Trinidad.*

The last case study in this section puts the spotlight on the immigrants themselves and the challenge of adjusting to their new homelands. What happened to those immigrants from Trinidad and other Caribbean countries, as well from Africa and Asia, who entered the UK after the Second World War?

A disturbing verdict?

There was a substantial influx of Asian and African-Caribbean immigrants into the UK during the 1950s and 1960s. A significant proportion were young adults who have since reared families and become grandparents. Data from the latest census have given the most detailed picture of ethnic Britain. The census form allowed people to describe themselves as mixed race, rather than having to choose white, Asian, black or Chinese ethnicity, while more questions were asked about people's homes, family set-up and overall health. People from ethnic minority groups now make up almost 8% (4.6 million) of the population, an increase of 53% over the 1991 census. Figure 24 shows the breakdown of the UK's ethnic population. The fact that 15% of that population declared themselves as being of 'mixed race' provides evidence of integration. So too does the fact that half of that ethnic population was born in the UK. It emphasises the crucial point that these people are part of the community and not foreigners.

It is a widely held belief that first-generation immigrants often suffer in their new homeland, particularly in terms of having to take low-paid jobs and occupy poor housing. In contrast, the offspring of immigrants (i.e. the next generation), since they are citizens of the country in which they were born, are seen as typically moving off the breadline and becoming better off. However, the results of the 2001 census challenge these beliefs. They show that black, Asian and other ethnic minorities are twice as likely to be unemployed, half as likely to own their home and run double the risk of poor health than white Britons. They also show that the proportion of Muslim children living in overcrowded housing is more than three times the national average, that they are twice as likely to live in a house with no central heating and that children from Pakistani and Bangladeshi families suffer twice as much ill-health as their white counterparts. Residential segregation on the basis of ethnicity, particularly in inner suburbs, remains the rule. Is that by choice or force of circumstance?

In short, the census results indicate that the gap between the richest and poorest in the UK has never been wider. More worrying is the fact that the widening gap has an ethnic dimension. One researcher has described the situation as follows:

> Some people from ethnic minorities are moving up and becoming wealthier, but for some children the future is bleak. They are getting the worst start in life and the problem becomes a vicious circle because of the correlation between poor housing, poor education and future criminality.

It is difficult to know exactly what needs to be done to ensure a fairer future for all. Successive governments have set up a range of programmes to tackle discrimination, poverty and inequalities, but they simply do not seem to be working. Instead, it looks as if generations of the UK's ethnic minorities will continue to grow up in deprivation.

However, all is not gloom; there is at least one ray of hope. A recently published survey has shown that a clear majority of people from all the main ethnic minorities identify themselves as British (Figure 25). Feelings of 'Britishness' are strongest among people of mixed race.

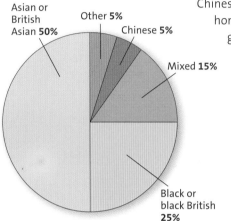

Asian or British Asian **50%**
Other **5%**
Chinese **5%**
Mixed **15%**
Black or black British **25%**

***Figure 24** The composition of the UK's ethnic population, 2001*

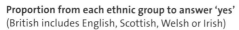

Proportion from each ethnic group to answer 'yes'
(British includes English, Scottish, Welsh or Irish)

Figure 25
'Do you feel British?'

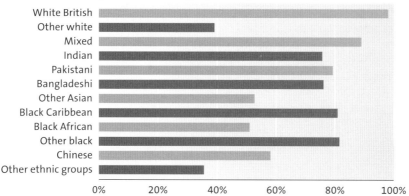

As a postscript, the point needs to be made that the UK today is not just home to three generations of New Commonwealth immigrant. As Figure 26 shows, the UK also continues to attract significant flows of immigrants from other parts of the world. Equally important, it illustrates the two-way nature of migration, with the UK also acting as a source.

This case study shows that, despite well-intentioned legislation, ethnic minorities in the UK still do not enjoy the same range of opportunities as the white majority. There are two related discussion issues:

- *What else might be done to ensure equality?*
- *Is the loss of ethnic identity the price that has to be paid for achieving equality and integration?*

10 Using case studies

Perception is an important aspect of migration. It arises in two different, but related, contexts. First, there is the perception of the migrants themselves, particularly of the push and pull factors impinging on them. In many cases, it is the perception and valuation of opportunities at the potential destination that is particularly important. Second, there is the perception of migrants by people living in the source and the host areas. What do they think about those who are leaving? Perhaps more critically, what do they think of the incomers? Here is a perception exercise for you to work on.

Question

(a) For each of the incoming groups shown in Figure 26:
  (i) put yourself in the place of one member and identify your likely perception of the upside and downside of a move to the UK
  (ii) try to identify the likely attitudes of those remaining behind towards these emigrants
(b) For each of the outgoing groups shown in Figure 26:
  (i) put yourself in the place of one member and identify your likely perception of the downside of living in the UK
  (ii) try to identify the likely attitudes of people in the host country towards this group of migrants
(c) Compare your responses to (a) and (b). Are there significant differences or similarities?

EU workers
- Freedom of movement within the EU encourages large flows, particularly from 'southern' member states (Portugal, Spain, Greece)
- Immigrants from the EU account for only 13% of UK immigration

Non-EU workers with permits
- Mainly from eastern Europe, particularly countries joining the EU in 2004
- 'Work-related' moves account for 26% of UK immigration
- 34% of all UK immigrants come from outside the EU and Commonwealth

Illegal immigrants
- Mainly employed in the black economy, including Chinese cocklers, Albanian prostitutes, east European farm gangs, etc.
- No official statistics; estimates put the flow in the order of tens of thousands

Asylum seekers
- Today mainly from the Middle East (Iraq, Iran, Afghanistan) and China
- Significant flows from Africa (Somalia, Zimbabwe, Sierra Leone)
- Influx from former Yugoslavia now easing
- Account for 15% of all UK immigrants

Family and friends
- Relatives of families already resident in the UK
- Includes the 'victims' of 'forced' marriages.
- 'Accompanying/joining' accounts for 18% of UK immigration
- New Commonwealth citizens are the largest single immigrant group (32%)

Students
- Mainly from Europe and Commonwealth countries for degree and English language courses
- 'Formal study' accounts for 18% of UK immigration

Returning ex-patriates
- Often due to death of partner, disillusionment, end of contract
- Significant numbers from within the EU (France and Spain)
- 22% of UK immigrants hold British passports

Overseas postings
- Mainly professional people and technicians taking up long-term contract work, often in LEDCs
- This category also includes the 'brain drain' to other MEDCs
- 30% of emigrant moves from the UK are 'work related'

Discontents
- People dissatisfied with their quality of life, the cost of living, job prospects, etc.
- France and Spain may be popular destinations
- There are no statistics; they are included in the 'other reasons' category accounting for 34% of all emigration

Refused asylum seekers
- Most of these are economic migrants who are 'removed' and sent back to homes in eastern and southeastern Europe, the Middle East and New Commonwealth countries
- Account for less than 5% of emigrants

Returnees
- Normally first-generation immigrants who wish to retire to their roots, having made sufficient money working in the UK
- They account for a significant proportion of the 16% of emigrants bound for Commonwealth countries

Students
- Includes 'gap year' students and undergraduates going to universities in North America and Old Commonwealth countries
- They account for around 5% of UK emigrants

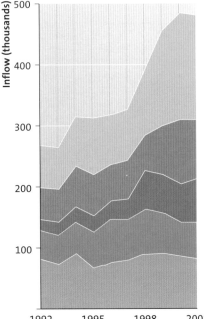

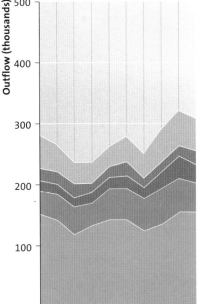

Key **Country of birth**
- Other
- New Commonwealth
- Old Commonwealth
- EU
- UK

Figure 26 UK total international migration, 1992–2001

11

Question

With reference to specific examples, assess some of the benefits and costs of international migration to both source and host countries under the sub-headings:
(a) environment
(b) economy
(c) society and culture

Guidance

You might tackle this question using the information contained in *Case studies 7–10*. Start by filling in a table such as the one below:

Source country	Host country
Benefits:	Benefits:
Costs:	Costs:

The content of the table should lead to the following assessments:
- In neither location is the balance between benefits and costs wholly one-sided.
- The balance generally tips in favour of costs in the source country.
- The balance generally inclines the other way in the host country.

Consumption

People are first and foremost consumers. The very survival of the human race requires basics such as air, water, space, food and shelter. Such survival needs are in effect **resources:**
- **natural** (e.g. water, soil)
- **human** (e.g. enterprise, education)
- **renewable** (e.g. solar energy, forests)
- **non-renewable** (e.g. fossil fuels, minerals)

The rate at which **resources** are consumed (**consumption**) is strongly influenced by two processes: **population growth** and **development**. We can think of population growth, resources and development as being involved in a pattern of triangular relationships that are reciprocal or two-way (Figure 27).

We can call this set of relationships the **consumption triangle**. Figure 27 shows that while population growth enriches human resources (such as labour and enterprise) it also raises the demand for, and the exploitation of, resources. A knock-on effect of this is to encourage economic growth and therefore development. Development may be expected to impact on population growth by reducing mortality (better healthcare) and encouraging in-migration (economic asylum seekers). A 'circuit' of relationships can also be detected going in the opposite

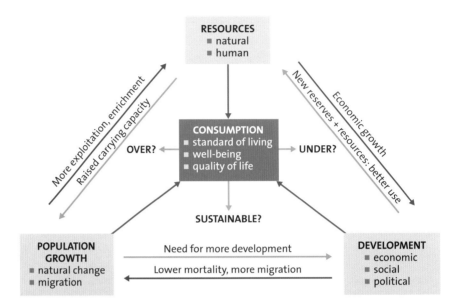

Figure 27
The consumption triangle

direction. For example, population growth creates a demand for development. More development increases resource exploitation. Equally, new technology (a part of development) could lead to more efficient use of resources, as well as to the discovery of new reserves and resources. A resource base enlarged in these ways will raise the carrying capacity of an area and therefore its ability to support more population growth.

Sitting in the middle of the triangle is **consumption**. While the rate of population growth is a critical factor here (more people mean more consumption), so too is the ability of development and the resource base to sustain rising levels of consumption. If that capability exists, there is the promise of rising levels of **human welfare**, as reflected in a range of measures such as **standard of living**, **well-being** and **quality of life**. If that capability does not exist, deterioration in the condition of a population seems inevitable. Therefore, it is the nature of the balance between resources, development and population growth that is of paramount importance. Broadly speaking, three different states of 'balance' may be recognised, each at two levels — national and individual:

- **Over-consumption** at a national level occurs when, and where, population growth races ahead of the current availability of resources and the present level of economic development. It involves low standards of living, and its symptoms include poverty, malnutrition, disease and environmental damage. However, at a personal level, over-consumption involves a totally different scenario. Among its more obvious symptoms are affluence, a profligate use of resources and obesity.

- **Under-consumption** at a national level is relatively rare today. It exists where resources and development could support a larger population without any serious depletion of known reserves. In stark contrast, under-consumption at an individual level is widespread. Its symptoms include those characteristics mentioned above in connection with national over-consumption, namely poverty, malnutrition, disease and short life expectancy.

Contemporary Case Studies

- **Sustainable consumption** is a much sought after state of balance, but one that could prove difficult to achieve nationally. It requires a stable population, innovative development and the consumption of renewable rather than non-renewable resources. It is a situation in which the living standards of future generations are not threatened by present consumption. However, as individuals, each of us can do a great deal to make our lifestyle more sustainable. Actions we can take include, among others, recycling, using public transport rather than the car and eating organic food.

Over the years, three main theories have been put forward about the relationship between population and resources (Figure 28). They can be summarised as follows:

- **Malthus** (1798) argued that food supply acts as a ceiling to population growth and that population growth takes place at a faster rate than the increase in food supply.
- **Boserup** (1965) argued that population growth stimulates increased food supply, either by improving the productivity of farming or by importing supplies.
- The **Club of Rome** (1972) predicted that if the current trends in population growth, industrialisation, food production, pollution and resource depletion were maintained, the limits to global population growth would be reached within 100 years. Population would then decline. Before that point, policies could be introduced to decrease fertility and therefore lower the rate of population increase.

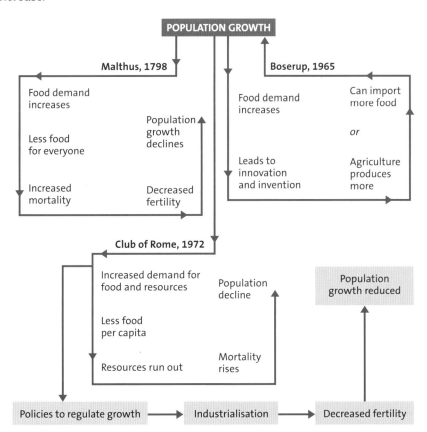

Figure 28
Three theories about the relationship between population growth and resources

Chile today

Chile is one of the most weirdly shaped countries in the world. It occupies a thin strip of land, rarely more than 200 km wide, which stretches 4640 km down the west coast of South America from north of the Tropic of Capricorn to Cape Horn (55°S). As a consequence, the climate ranges from hot desert in the north to cool temperate in the south. Southern Chile is one of the world's wettest and stormiest regions. Only 2% of Chile's land area is suitable for growing crops, 17% for livestock and 11% for forestry. The remaining 70% is covered by deserts or mountains (the Andes dominate) and is considered unproductive for agriculture, but suitable for mining.

With a population of just over 15 million living at a mean density of 20 persons per km^2, Chile ranks as one of the most sparsely populated countries. Does this necessarily mean, however, that it is in the enviable position of being a country of underpopulation and under-consumption?

Distribution of population

If we look at the distribution of population in Chile, we might readily conclude that the central regions are overpopulated (Figure 29). In the two contiguous regions of Santiago and Valparaiso, just over half of Chile's population is living on a mere 4% of the land area, at a mean density of 240 persons per km^2 (six times the national average). Of the various climates that Chile has to offer, the Mediterranean type found here in the middle of the country is the most attractive. However, high levels of atmospheric and water pollution indicate that the concentration of so many people and so much of the country's economic activity in the central area is seriously degrading the environment. Air pollutants become trapped within the Central Valley cut deep into the Andes, making Santiago one of the world's most polluted cities.

If we look outside the central area and focus instead on the remainder of the country, with its mean density of 10 persons per km^2, we might conclude that Chile is, after all, an under-populated country.

Be careful here. The key to understanding overpopulation is not population density but the numbers of people in an area relative to its resources and the capacity of the environment to sustain human activities. A country becomes overpopulated when its population cannot be maintained without the rapid depletion of its non-renewable resources and without degrading the capacity of the environment to support the population. In short, if the current human occupants of a country are clearly degrading its long-term carrying capacity, that country is overpopulated.

Chile and the global economy

Like most of the rest of the world, Chile is being drawn into the global economy. It now meets a significant proportion of the global demand for copper and nitrates. These are the chief exports (accounting for around half of all exports by value in 2001). Other exported minerals include heavy bitumen and iron ore. This dependence on selling non-renewable resources does not bode well for a sustainable future. There is also much environmental damage to take into account.

The situation with respect to renewable resources is not much better. Large exports of fish and crustaceans (fresh and prepared) are depleting the stocks of Chile's once-rich waters. Still worse is the large-scale deforestation caused by big exports of timber, wood pulp and paper. While the rising production and export of fruit and wine may look to be a promising development, it has required taking over land once used for subsistence purposes (both crops and meat). Remember that agricultural land is a scarce resource in Chile. So while Chile exports food and drink products worth $5 billion, its imports in the same category amount to $1 billion. The irony is that Chile is quite capable of producing many of those imported products — indeed, it used to do so. It seems perverse that Chile is not doing more to meet its increasing food needs by exploiting its own range of climates. In theory, this climatic range should allow it to produce almost anything, from tropical crops to cool temperate livestock.

The future?

Two indicators suggest that Chile may well be living dangerously. Its consumption pressure is currently put at 3.26 (three times the global average) and its ecological footprint at 3.2 (one and a half times the global average).

There is no doubt that since the political troubles of the 1970s and 1980s, the economy of Chile and the standard of living of its people have improved. It is now recognised as a 'middle-income' economy. However, both these achievements have involved much higher levels of consumption. There are serious warnings that Chile may be reaching the critical thresholds of over-consumption and overpopulation, particularly given the current population growth rate of 1.7% per annum.

This case study underlines the following points:

- *Population density is a poor indicator of overpopulation.*
- *Globalisation raises consumption and often this will involve a country exploiting its resources for the benefit of foreign rather than domestic consumers.*
- *Renewable resources, such as fish stocks and forests, are vulnerable to over-exploitation.*
- *Overpopulation exists where there is serious depletion of resources and environmental degradation.*

Figure 29 Chile: the distribution of population, by region, 2002

> ■ The speed at which a country reaches the threshold of overpopulation is influenced by the population growth rate.
>
> ■ The speed at which a country reaches the threshold of over-consumption is influenced by the economic growth rate.

The next case study involves a very literal use of the word 'consumption' and shifts the focus from populations towards people as individuals.

Case study 12 · THE FAT OF THE LAND

Over-consumption in the USA

US government figures show that 20% of Americans are obese, up from 12.5% 10 years ago. The USA is suffering from an 'obesity epidemic' that is quite simply due to an overdose of over-consumption (Figure 30). It is not only that people are eating too much, they are eating the wrong things: too much fast food, too much fat, too little fresh fruit and vegetables. Not taking enough exercise compounds the problem. It is difficult to know why so many are overeating when the health risks are widely understood. Alzheimer's disease, breast and cervical cancer have recently been added to the already staggering list of diseases associated with obesity. One US expert has suggested that obesity 'is a complex issue…recent world events and the threat of terrorism have made people feel insecure. They are turning to food for comfort and caring less about the consequences of unhealthy eating'.

Figure 30
Obesity in the USA

Topfoto

The 'obesity epidemic' is having all sorts of repercussions in addition to the adverse impact on health and mortality rates. For example, fat is becoming a corporate issue. It is accepted as a fact of modern living; there is now less stigma attached to being fat.

Bigger and bulging consumers pose both challenges and opportunities. On the one hand, they put greater strain on existing products. This means that manufacturers must spend more to stop things, such as chairs, beds and seats in public places, from falling apart. On the other hand, bigger consumers represent a new niche market for 'super-size' products. Some airlines and public transport providers are starting to cater for bigger bottoms. Manufacturers and retailers are selling more ranges of 'plus-size' clothing.

Gone are the days when fat people were the targets of ridicule and humour. Today, they are targets of some sympathy and big business. However, that change does not make the wider picture any less obscene — while some people stuff themselves to an early death, many continue to starve to an equally early death. Remember, too, that obesity is not just the USA's problem; the epidemic is spreading to many MEDCs, including the UK.

This case study reminds us that over-consumption has a serious downside that includes needless use of resources, diminished health, more spending on healthcare, and premature death. In its turn, this downside creates opportunities for the business world. Seizing those opportunities leads to further consumption.

The final case study in this part may at first sight seem to be a little peripheral to population geography. However, the point needs to be made that population growth and the development process literally 'drive' the ever-increasing number of motor vehicles on the global road network. The motor vehicle is one of the main modes of migration. It is the epitome of consumption. It is rapidly eroding the global stock of non-renewable resources and causes untold damage to people and the environment.

THE GLOBAL LOVE AFFAIR WITH THE CAR

Case study 13

Consumption with real costs

When it comes to discussing human consumption, perhaps no commodity or product excites such strong feelings as the motor vehicle. On the one hand, it is a vital part of modern life, being particularly crucial to economic development. On the other, it consumes an immense amount of oil (an increasingly scarce non-renewable resource), and it is a major polluter of the atmosphere and therefore a prime contributor to global warming.

Here are some facts about motor vehicles:

- 3000 people die on the world's roads every day; 3 million are injured in road accidents in Europe each year.
- Cars produce 15% of global greenhouse emissions; in MEDCs the figure is 20%.
- Europeans make 62% of their journeys by car; Americans, 86%.
- There are more cars than adults in the USA, largely due to the rise in sports utility vehicles as a driver's second or third car.
- Toxic emissions from modern cars are 30 times lower than those from cars built 20 years ago.
- Ten cars today make the same volume of noise as one did, 30 years ago.
- 40% of British drivers express concern about their car's impact on the environment, compared with 30% of Americans.

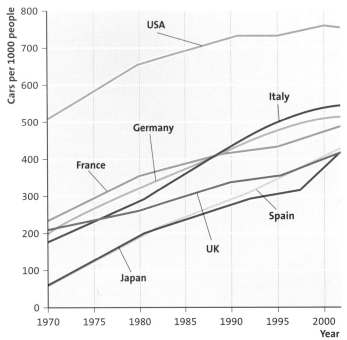

Figure 31
The growth in car ownership in selected countries, 1970–2000

Transportation of all types accounts for more than a quarter of the world's commercial energy use. Motor vehicles consume nearly 80% of all transport-related energy. They have brought enormous economic and social benefits. However, their widespread use has environmental and economic costs that have risen sharply over the past few decades, as vehicle numbers have increased. Motor vehicles are a major source of urban air pollution and greenhouse gas emissions. They also represent a serious threat to the economic security of many nations because of the need to import oil to fuel them.

In 1950, there were around 70 million motor vehicles on the world's roads. Today that number is almost ten times greater. If this kind of linear growth continues, by the year 2025 the number will be approaching 1.5 billion. Clearly, there is enormous potential for further increase in vehicle use. Per capita car ownership is high in MEDCs, but it is still low in LEDCs. The growth potential is especially great in the developing economies of Asia. For example, in China there are only about eight vehicles per 1000 persons; in India only seven per 1000. By contrast, there are about 760 cars per 1000 persons in the USA and 410 per 1000 in Japan and the UK (Figure 31). The Italians own proportionally more cars than any other Europeans, with 540 per 1000.

With the exception of Denmark, which has levied a punitive purchase tax on cars, there is as yet little evidence of governments making serious efforts to reduce car ownership. Indeed, governments have a vested interest in the continued use of motor vehicles. They are mindful of:

- the jobs involved in the manufacture, servicing and sale of vehicles
- the huge tax revenues from the sale of fuel
- the ease and convenience of the motor vehicle in terms of assembly, distribution and movement generally

A global future of more people and more development means that there will be no reduction in either the demand for motor vehicles or in their environmental and social costs. Designing a new generation of resource-efficient, environmentally friendly vehicles is one of the most challenging technological problems facing the industrialised world. Most major vehicle manufacturers are beginning to respond to this challenge. They need to hurry up, before the oil runs out.

This case study presents an example of consumption where demand is being driven by a number of powerful factors: population growth, development and modern lifestyles. The love affair with the motor car seems to have made the global community either blind to, or extremely tolerant of, its undoubted costs.

Contemporary Case Studies

12

Question

Study Figure 32, which shows the three different relationships between population and resources.

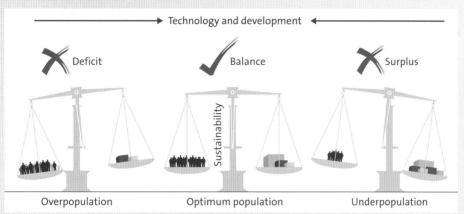

Figure 32
The three different relationships between population and resources

With reference to examples, evaluate the range of strategies available for countries to achieve the goal of an optimum population.

Guidance

The diagram suggests that there are three main courses of action:

■ control population size, i.e. reduce overpopulation
■ control resource use, i.e. reduce consumption
■ use technology and development to bring population and resources into a sustainable balance

To reduce overpopulation, the main strategy is to target and reduce birth rates. Appropriate examples to use are China (reasonably successful — *Case study 17*) and India (distinctly unsuccessful). Another possible strategy is to change the distribution of population by encouraging people to move from congested to underpopulated countries or areas (Indonesia — *Case study 27*).

To reduce consumption, the aim is to steer away as much as possible from the use of non-renewable resources and to encourage recycling and conservation (the Netherlands is a leading light in this respect).

The trouble with both technological progress and development is that they have an in-built tendency to raise the consumption of resources (*Case study 13*). There are few, if any, countries that have as yet really harnessed new technology to help reduce resource consumption and yet allow living standards to rise. Chile (*Case study 11*) is one country where the chances of achieving this sustainable balance are good, provided that there is a planned redistribution of the population and an investment of new technology in the opening up of regions that are presently underdeveloped.

Part 3

Illustrating population policies and issues

Policy refers to those actions of governments that try to control, manage or influence, in this instance, specific aspects of population and migration. To a lesser extent, decision makers in the private sector may also have some impact. For example, industrialists, in their recruitment of labour, may well encourage migration; so too developers, particularly through housing developments. However, the hand of public sector or government intervention can be seen in all of the four concepts covered so far (Figure 33). Governments seek to manipulate:

- the distribution of population
- population change
- population migration
- consumption, by actively encouraging the development process or more resource use

Figure 33
Some examples of population policies

Distribution
- Settling areas for political reasons (e.g. the West Bank, Israel)
- Opening up 'new' or 'difficult' areas (e.g. Dubai)
- Halting the downward spiral in declining regions (e.g. rural Spain)
- Relieving over-crowding (e.g. UK's New Towns programme)

Change
- Implementing birth control (e.g. China)
- Raising rates of population growth (e.g. Romania)
- Meeting the needs of 'greying' populations (e.g. the UK)
- Instigating demographic change through development (e.g. Asian NICs)

POLICY

Migration
- Encouraging emigration (e.g. Kurdistan)
- Welcoming immigrants (e.g. Australia)
- Exploiting immigrants (e.g. Vietnamese in Germany)
- Imposing immigration controls (e.g. the UK and asylum seekers)

Consumption
- Raising food production (e.g. FAO)
- Moving towards sustainability (e.g. Kyoto Protocol)
- Relieving overpopulation (e.g. Indonesia's transmigration programme)
- Confronting underpopulation (e.g. Italy)

Eight case studies follow to illustrate these policy 'areas'. These are followed by a section that looks at four global issues, each relevant to one of the four key concepts of distribution, change, migration and consumption.

Distribution

There are a number of different circumstances that might encourage governments actively to alter the distribution of population. These range from trying to achieve a spread of population that matches the distribution of resources and economic opportunities, to extending the nation's territorial grip.

JEWISH SETTLEMENTS IN THE WEST BANK

Case study **14**

Population policy with a political purpose

The state of Israel came into being in 1948, when the UN decided to split what had been Palestine and to make a Jewish and an Arab state. Needless to say, the Arabs resisted the partition and have continued to do so ever since. This part of the Middle East is one of the world's most persistent and violent trouble spots. To add to the tension, in 1967 Israel occupied a sizeable tract of the much-reduced Palestine (the so-called West Bank, roughly the size of Norfolk), mainly on the grounds that it would make Israel's borders more secure against attacks from its Arab neighbours. Since then, successive governments have pursued a policy of consolidating the West Bank as an integral part of Israel.

Three related strategies have been involved in the implementation of this policy. First, the Israeli government has sought to 'persuade' some of the Arabs to move away and settle in what is left of Palestine. Second, it has tried to concentrate the remaining Arabs in designated settlements and refugee camps. Recently, a British ambassador to Israel described the West Bank as 'the largest detention camp in the world'. Third, between the scattered areas of Arab settlement, the Israeli government has encouraged the building of well over 100 new settlements occupied exclusively by Israelis (many of whom are recent émigrés from eastern Europe) (Figure 34). The construction of some of these has involved the demolition of Arab settlements. Some of the new settlements have not been officially sanctioned.

The net outcome of these actions or policies has been to:
- raise the population density
- expand the settlement network
- concentrate the Arab people into fragmented areas
- increase the Jewish component in the population

Today, the population of the West Bank is around 2.2 million, with the Jewish element amounting to only 17%. An added difficulty in moving towards a Jewish majority in the West Bank is the much higher fertility of the Arabs — 34 per 1000, compared with 20 per 1000 for the Israelis.

While Israel may feel more secure with the 'converted' West Bank as part of its territory, the continuing terrorist activity, particularly the suicide bombings, must remind the Israelis that their occupation and settlement of it may not be too secure. Indeed, Israel has intermittently indicated a willingness to discuss the permanent

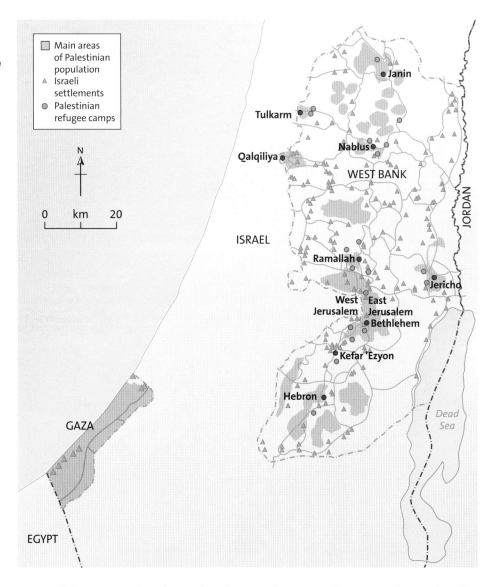

Figure 34
Established Jewish settlements in the West Bank and Gaza

status of the West Bank and Gaza (a strip of territory taken from Egypt in 1956) and to dismantle some of the 'illegal' Jewish outposts. However, so long as Jewish settlement continues, the hopes of a settled future for the West Bank remain remote.

This case study provides an illustration of how government policy can affect both the distribution and the composition of an area's population. In this instance, the reasons for doing so are political and mainly to do with security. Whether the policy will ever achieve a permanent solution remains in doubt.

Case study 15   PEOPLING AND GREENING THE DESERT

The rise of Dubai

Dubai is one of seven small states, fronting the Arabian Gulf, that make up the United Arab Emirates (UAE). One hundred and fifty years ago, its territory of 3900 km² was little more than barren, sandy desert. The only settlement was a small port of a thousand or so inhabitants, located on the banks of Dubai Creek, and relying mainly on fishing, the pearl industry and overseas trade. Today, Dubai is an amazingly prosperous state. In addition to oil revenues, it thrives on a mix of corporate business (attracted there by zero taxation), re-exporting (helped by low duties) and tourism (thanks to its beautiful beaches, warm seas, sumptuous hotels and duty-free shopping). The population has risen to nearly 1 million, most of whom live in the city itself. Dubai city is essentially linear, running parallel to the coast from Deira (old Dubai) in the direction of Abu Dhabi (another emirate). The built-up area has some of the most spectacular architecture in the world, notably the futuristic skyscraper office blocks and hotels, as well as many modern shopping malls (Figure 35).

It has been said that 'the story of Dubai reads like a rags-to-riches tale'. It is hard to imagine anywhere else in the world that has developed at such a pace, in such a short time. The desert sands are being hidden, not just by the built-up area, but by grass and palm trees. One might say that the desert is beginning to 'blossom'. Much of Dubai's success has to be credited to the remarkable vision, enterprise and leadership of the ruling Maktoum family, who have been the policy- and decision-makers.

Jochen Tack/Still Pictures

Figure 35
Some of Dubai's dramatic architecture

The building and running of modern Dubai has involved, and continues to involve, two distinct waves of immigration. Construction has required the recruitment of technicians and professionals from western Europe and North America, and also of manual labour, mainly from Africa and the Indian subcontinent. The latter regions also supply most of the labour required now to fill a wide range of jobs, from shop-counter and hotel posts to domestic servants for wealthy Arab families and the general running of public services. This heavy reliance on immigrants is reflected in the composition of Dubai's population today. Only 18% of the population is made up of UAE nationals; 13% comes from neighbouring Arab states, while 65% are expatriates from other parts of Asia. In order to gain entry into the country, an immigrant worker must be sponsored. Contracts of employment are for up to 2 years. However, an immigrant may only work for the sponsoring employer. For these reasons, there is a remarkable turnover in Dubai's population. Sadly, most of the immigrant workers are treated by the Dubai-ites as second-class citizens. If a worker earns less than 3000 dirhams (approximately £500) a month, they are not permitted to bring their spouse and family into the country.

If one had to name one resource, other than oil, that has been crucial to the rise of modern Dubai, it would be water. In a virtually rain-free part of the world, Dubai has had to look to the desalination of sea water to provide most of the 700 000 m^3 of water needed each day. Apart from human consumption, water is needed for the constant irrigation of golf courses, grass verges and palm trees. Water lost through high rates of evaporation from the many fountains that punctuate the built-up area also needs to be replaced. Small wonder that Dubai has been at the cutting edge of new desalination technology. It has pioneered the building of some of the largest desalination stations in the world.

Resource consumption in Dubai involves much more than just oil and sea water. The extravagance of nearly 1 million inhabitants and over 3 million tourists a year is immense. The demands range from electricity (much of it to help Dubai keep cool in sweltering temperatures) and high-grade building materials to top-quality food and the latest in electronic gadgetry.

This case study makes the basic point that even the most hostile of environments is capable of settlement and development. Success or failure depends largely on:
- *sound decision making and management*
- *the use of modern technology*
- *the availability of much capital*

13 *Using case studies*

Question

'The impact of the physical environment on the distribution of population is greater in LEDCs than in MEDCs.' Discuss this statement.

Guidance

An answer to this question might make use of *Case studies 2* and *3*. Since the latter may be argued as an exception to the rule, it would raise the overall quality of the answer if we were able to introduce another MEDC example, but one that conforms. You could choose *Case study 14* or even *Case study 15* for this purpose.

You might start by pointing out that the three countries studied differ not just in terms of development but also in terms of their physical environments. Japan covers a large latitudinal range, from cool temperate to subtropical; three-quarters of it is inaccessible upland. The relatively small amount of lowland is highly fragmented and occurs mainly around the coast. By contrast, Ethiopia shows great altitudinal range, which covers roughly the same range of climates as in Japan. Israel is a small country. Extensive tracts of arid and semi-arid terrain lie between the coastal belt in the west and the River Jordan in the east.

Having described the physical background, we should now move on to the patterns of population distribution. For all its economic strength and technological know-how, the distribution of population in Japan is very much governed by the distribution of lowlands. Much of the upland interior of the country is inaccessible and of little value except for its forests and for some forms of tourism and recreation. In Ethiopia, with its poverty and little know-how, the population tends to avoid the extreme altitudes (because they are either too cold or too hot and arid). In Israel, much has been done to 'green' the arid areas and to reduce the early concentration of settlements along the coast and Jordan valley.

Thus, in two of the countries, despite the huge development gap between them, the physical environment exerts a potent influence on the distribution of population. As such, Japan refutes the generalisation, but Israel supports it in that modern technology has been used to largely overcome a distinctly hostile physical environment and to even out the distribution of population.

Contemporary Case Studies

Change

While the rate of population growth has for a long time been a major concern, recent history is peppered with examples of governments wishing to boost their populations. This was the case in Europe after the First and Second World Wars. There was a desperate need to make up for those killed during the fighting and to restore the labour force to the required level. More recently, some MEDCs have become alarmed by the rapid 'greying' of their populations (*Case study 22*) and by the depopulation of rural areas (*Case study 8*). Basically, there are two ways to boost a population. One is to encourage couples to have more children. The second is to encourage immigration, particularly of people in the reproductive age range. The latter option offers a 'quicker fix' and seems to be more favoured at the present time, as illustrated by Spain (*Case study 8*) and Australia (*Case study 25*).

After the Second World War, France and the UK took both options, but have since gone rather cool on immigration (*Case studies 10* and *28*). Perhaps the most draconian example of taking the first option is provided by Romania during the period 1966–90, as explained below.

BOOSTING POPULATION GROWTH

Case study **16**

Romanian attempts

In 1944, Romania was overrun by troops from the USSR and from then until 1989 it survived as a socialist state within the Soviet bloc. For much of that time, Romania experienced a very low rate of natural increase. Indeed, in 1947 it had fallen to as low as 1.4 per 1000. The reasons for this were mainly:

- the pillaging of the country's food production by the Red Army
- the persistently high death rate, boosted by widespread famine
- the availability of abortion on demand
- urbanisation and associated poor, cramped housing
- changes in society, with females staying in school longer and also delaying having children

Soon the disturbing consequences of near to zero population growth became apparent to those planning the economic development of the country — there would not be sufficient labour for the planned industrialisation. So in 1966 the government took a number of steps to boost the rate of natural increase by focusing on the birth rate. These steps included:

- prohibiting abortion on demand; abortion was permitted only in a limited number of clearly defined circumstances
- imposing a special income tax on men and women who remained childless after the age of 25, whether married or single
- stopping the import and sale of contraceptives
- making divorce much more difficult

The above were the 'sticks' of the policy; the 'carrots' included:

- increases in family allowances paid by the state

- a monetary reward given to mothers, beginning with the birth of the third child
- a reduction of 30% in the income tax rate for parents of three or more children

The impact of these pro-natalist policies was immediate. The number of live births in 1967 was almost double the number in the previous year. However, the increase was short-lived and by 1983 the birth rate had slipped to 14.3 per 1000. Even more draconian measures were then taken:

- The legal age of marriage was lowered to 15.
- Additional taxes were levied on childless individuals over 25 years of age.
- All women of child-bearing age were subject to monthly medical examination to identify pregnancies in the earliest stages and to ensure that these pregnancies came to term.
- Miscarriages were investigated and people involved in illegal abortions were prosecuted, with prison terms of 1 year for the women concerned and 5 years for doctors and other medical personnel performing the procedure.
- To qualify for an abortion, a woman had to have had five children, with all five still under her care, or be over 45 years of age.
- More monthly allowances were made for families with children, with bonuses given for the birth of the second as well as the third child.
- Mass media programmes extolled the virtue of large families and proclaimed that it was the patriotic and moral duty of all citizens to procreate.

Despite this next round of extreme measures, there was only a slight increase in the birth rate to 16 per 1000 in 1986. Even more worrying for the government was the fact that the cost of the incentives and their 'policing' equalled half the amount budgeted for defence.

The failure of all these measures to have the desired effect has to be put down to the failure of the government to create the right socio-economic conditions. For example, a woman's double burden of childcare and full-time work was not eased by the poor availability of consumer products (such as washing machines and prepared foods) that save time and labour in the home. Rationing of basic food items, shortages of other essential household goods, and overcrowded and substandard housing were other good reasons why people preferred not to have children. No wonder that so many women risked all to have illegal abortions, which at one time could be performed in the 'back streets' in exchange for a packet of cigarettes.

Mercifully, this tale of horror came to an end in 1989 with the toppling of the communist government. Looking back, it seems strange that the government did not focus more of its efforts on lowering the mortality rate, particularly the rate of infant mortality. Perhaps this would have been a more effective way of raising the rate of natural increase.

This case study is about what must go down in history as possibly the most extreme pro-natalist policy ever attempted. The fact that it did not succeed highlights the point that decisions about having, or not having, children are highly personal. When it comes to governments attempting to change those decisions, gentle persuasion is more likely to succeed than coercion, and persuasion will only work if the conditions are right.

Just as Romania's pro-natal policies seem extreme, so too does China's anti-natal (or one-child) policy.

Contemporary Case Studies

CHINA'S ONE-CHILD POLICY

The backlash

Much has been written about China's attempts to curb the rate of growth of its immense population. In 1980 it introduced the draconian one-child policy. Except in a few particular circumstances, for nearly 20 years no couple was allowed to have more than one child. Those who did were penalised in various ways. Since 1999, there has been some relaxation of the policy, particularly in rural areas. However, the impacts of the original policy are evident in:

- a reduced birth rate (from 31 to 17 per 1000)
- a reduced growth rate (from 2.4% to 0.9% per year)

Even so, the total population has increased from 996 to 1266 million. So has the one-child policy really been a success? The answer is partly so. It has put the brake on population growth and that brake will become progressively more effective as the cutback in children works its way up the population pyramid (Figure 36). A number of demographic, unforeseen backlashes are already apparent:

- **uneven implementation** — the policy seems to have been implemented more effectively in towns and cities than in rural areas
- **negative growth** is now being experienced is some cities (e.g. Shanghai), where the policy may have been too effective
- **sex-selective abortion** — the traditional wish of fathers to produce a son remains strong; the sex ratio now stands at 116 males to every 100 females
- **increased divorce rate** — divorce is used by men as a means of ensuring a male heir
- **'little emperor' syndrome** — parents 'spoil' their 'one-boy' child, who as a result may become obese, demanding and delinquent

Figure 36
China's population pyramid, 2000

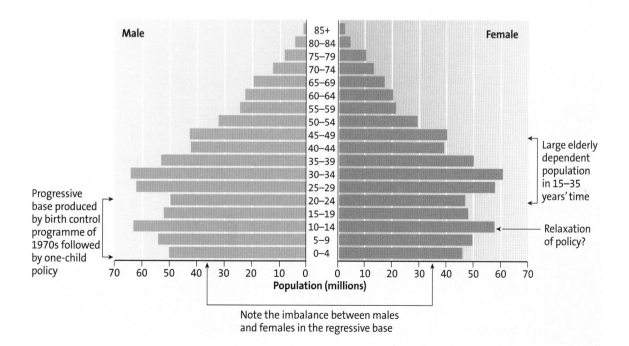

Figure 37
Chinese children in the classroom. How many of the boys will be able to marry when they are older?

- **shift in dependency** — while fewer children lower dependency at one end of the age spectrum, they raise it at the other end; the bulge in the middle of the present population pyramid (Figure 36) will soon result in more old people being supported by a reduced working population
- **gender imbalance** — created by the preference for a male child, this promises some serious social problems. For example, due to the increasing shortage of women of marrying age, bride bartering or kidnapping has already become commonplace in rural areas. Prostitution and the trade in sex slaves are now rife in the cities. The growing population of unattached 'little emperors' (Figure 37), particularly in the cities, could easily spark social instability and a rise in crime. Another possible solution might be to encourage the immigration of women from other countries.

The bottom-line question to this issue of gender imbalance is: why do Chinese parents continue to show a strong preference for boys over girls? In India, the explanation lies in the dowry tradition, but that tradition does not exist in China. Instead, it looks as if the reason may lie in the absence of any pension system and the belief that it is sons, not daughters, who can support parents in their twilight years. In 2004, the government recognised the problem and introduced its 'Care for Girls' plan. Farming families are to be offered incentives to stop the abortion of female fetuses. Girls will be exempted from school fees, while their parents will benefit from tax breaks, free insurance and extra housing.

Further down the line, it looks as if **labour shortages** will be another unwanted consequence. At the moment, unemployment prevails in most cities, a problem made worse by large-scale rural–urban migration. However, that will change as the reduced cohorts of children become part of the productive population. Although not directly linked to the one-child policy, it is worth noting that the food security of China is being threatened, as urban areas expand onto scarce arable land.

Even further down the line (say in 50 years) it is possible that, despite a relaxation of the policy, China's population may contract to 800 million. The population pyramid will then look like those of MEDCs that are entering Stage 5 of the demographic transition model. Is that what the policy makers in China really want?

This tough policy has done much to defuse the population explosion that so threatened China in the third quarter of the twentieth century. However, the case study also illustrates two vital points:

- *A policy aimed at one objective can often have unwanted and unforeseen impacts.*
- *A policy that prevails for too long can become counterproductive — such is the dynamic nature of all populations.*

Contemporary Case Studies

Question

Suggest and consider the extent to which the following factors are crucial to a successful birth control programme:

- accessible and affordable contraception
- female empowerment
- education
- religion
- economic incentives
- political persuasion

Guidance

Answering this question requires the rereading of not just the Romania and China case studies; you should also take a look at *Case studies 5* and *6*.

Migration

It is no exaggeration to say that today international migration is a major global issue. Increasingly, governments at both ends of the migration pathway feel the need to intervene, mainly to put the brake on either outflows from major source countries or inflows to popular host countries, particularly the latter. Migration policies often have sinister political motives (*Case study 18*). Equally, many governments are finding it increasingly difficult to curb the forces of economic globalisation that are so encouraging to international migration.

THE KURDS

Case study **18**

A nation of refugees

The traditional homeland of the Kurds straddles the mountainous area where the borders of four nations (Iran, Iraq, Syria and Turkey) now converge (Figure 38). Kurds in this region currently number around 25 million. They have never been allowed to establish an independent state and have suffered much persecution.

In Iraq, the Kurds suffered greatly at the hands of Saddam Hussein and his regime. During the 1980s, this regime 'eliminated' nearly 250000 Kurds. For example, in March 1988, 5000 Kurds died in a nerve-gas attack on the town of Halabja. After the Gulf War in 1991, the USA and its allies sought to establish a safe haven in the northern part of the country for the remaining 4 million Kurds. Not surprisingly, during the invasion of Iraq in 2003, the Kurds joined US and British forces in trying to rid the country of Saddam Hussein.

Turkey, which occupies over half of Kurdistan and accounts for half the Kurdish population, has been trying to beat down a determined Kurdish liberation struggle for some 20 years. Up to a million Kurds have been displaced from their homelands on the pretext that the areas were needed for security reasons or for dam construction. During the 2003 invasion of Iraq, the Turks were very nervous that the Kurds might succeed in pressing their claim for an independent kingdom.

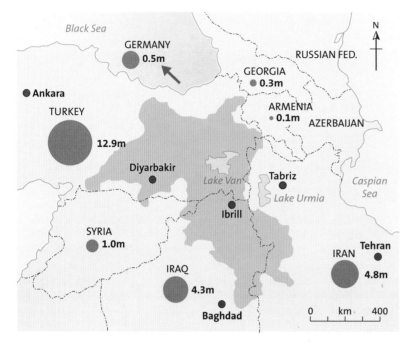

Legend:
Kurdish areas
Kurdish population in millions

The situation of the 4.8 million Kurds in Iran and the 1 million in Syria is not much better. Iran made use of the Kurds in the war with Iraq during the 1980s. Like Turkey, both Iran and Syria are reluctant to support any move for an independent Kurdistan, mainly because the territory is rich in water and oil resources.

Thus it is that the Kurds today find that they are, in a sense, 'refugees' in their own countries. It is common policy of the countries occupying Kurdistan to suppress the Kurds. Various strategies have been used including:

Figure 38
Kurdistan

- keeping the areas economically weak by curbing development
- denying human rights and access to education and healthcare
- 'persuading' Kurds to settle elsewhere

In short, there has been **ethnic cleansing**. The outcome is that Kurdistan has become a land fast losing many of its more enterprising people. There is considerable emigration.

The official migration statistics make no reference to Kurds, but most of the asylum seekers recorded by the United Nations High Commissioner for Refugees in 2002 as being from Iran, Iraq and Turkey (Table 5) are likely to have come from Kurdish areas. Most have headed either for Australia or the EU, particularly for Germany and the UK. However, the outflow of Kurds from their homeland has been occurring for decades, so the Kurdish presence abroad must now run into millions. The welcome they receive is often far from warm, as indicated by the cartoon (Figure 39), which was prompted by an outbreak of violence against Kurdish asylum seekers in Glasgow. On top of this, most face a future of poverty. The lot of the Kurds should prick the conscience of the world — unwanted in their homeland and rarely wanted wherever else they go.

Country of origin	Number of refugees
Iran	107 649
Iraq	581 233
Syria	8 225
Turkey	87 550

Table 5 Refugees and asylum seekers from four Middle East countries, 2002

Points raised by this case study:

- *The failure of political boundaries to recognise the existence and rights of an ethnic group can lead to huge human costs.*
- *The forces of push and pull are not always balanced. The push can be so powerful that it overrides the strength of the negative factors at the intended destination.*

The next case study tells of a situation not well known in the UK. What was happening in East Germany when the remarkable economic boom in West Germany was attracting a large influx of Turkish workers?

Figure 39
'Freedom at last'

©Richard Willson/The Times 2001

VIETNAMESE IMMIGRANTS IN GERMANY

Case study 19

Victims of the falling Wall?

The Vietnam War (1957–75) was long and highly destructive. It was essentially a conflict between capitalist South Vietnam (backed by the USA) and communist North Vietnam (backed by China). It was part of the so-called 'Cold War'. The destruction and economic dislocation it caused were so great that, for many years after the end of the war and the reunification of the country, tens of thousands of people tried desperately to escape. Perhaps the most publicised aspect of this flight from Vietnam were the so-called 'boat people' who took to the high seas, risking their lives in small craft as they tried to find a new home in either southeast Asia (particularly Hong Kong) or Australia (Figure 40).

As part of a package of international aid, the former East German government offered to take in an agreed number of Vietnamese people. The idea was that workers and their families would move to East Germany for a while (up to 10 years) and learn new skills, particularly in the textile industry. Once the skills had been acquired, they would return home and so contribute to the economic recovery of Vietnam. However, East Germany would also benefit from the labour provided by such immigrants. By 1989, there were 60 000 Vietnamese living in East Germany. They had become the largest group of foreign residents, accounting for 31% of the total. For example, in 1987 alone, the Karl-Marx-Stadt (Chemnitz) district welcomed over 5000 Vietnamese immigrants. Most were put to work in factories making clothing, umbrellas and cosmetics. They were well looked after by the local people, being provided with food, clothing and special housing.

Figure 40
Boat people escaping
from Vietnam

All went well until 1990, when the Wall that had separated East and West Germany for more than 40 years was removed and the two countries were reunited as a single state. Some of the Vietnamese people returned home as agreed. However, in the general confusion, many abandoned the scheme. Some seized the opportunity to set up their own businesses, such as snack bars, greengrocers and textile merchants, often trading in Vietnamese products. This broke the terms of their contracts and so they became illegal immigrants. At the same time, more illegal immigrants slipped into the country.

In 1995, there were an estimated 100 000 Vietnamese people in Germany. They formed fairly tight communities, tended to marry among themselves and produced children with strong affiliations to Vietnam. In the same year, the government announced that it would begin forcibly to repatriate some 40 000 Vietnamese who were regarded as illegal entrants. The Vietnamese government agreed to accept them back in exchange for a $130 million aid package and another $65 million to encourage German companies to invest in Vietnam. That worked out at about $5000 for each Vietnamese person sent home.

The question that springs to mind is: why was the German government so keen to send the Vietnamese back? Apart from the alleged involvement of some in cigarette smuggling, they had served the country well. There were a number of related factors moulding the decision:

- The economy of the reunified Germany was struggling and there was rising unemployment.
- Former West Germany had taken in thousands of guest workers (many from Turkey) during the boom days of previous decades, so there was already a supply of cheap, unskilled labour in that part of Germany. The Vietnamese were not needed there.
- Germany was in the process of having to readmit some 500 000 ethnic Germans who had been deported to Kazakhstan during the Second World War by Soviet forces. It was recognised that these people had the right to return to Germany and that they would need jobs.

- At a time of job shortages, it was felt that Germans should take priority over immigrants.
- The economy could not now provide for the consumption needs of large numbers of unemployed people and their dependants.
- Higher birth rates among immigrant groups raised concerns about Germany becoming too multiethnic a society.

As a postscript to this case study, it should be pointed out that many Vietnamese people still volunteer to work abroad. For example, there are currently 72 000 employed in Malaysia, Japan, Taiwan and South Korea. What also needs to be understood is the two-edged impact of these worker movements on the villages they leave behind. On the one hand, the villages are deprived of farm labour; on the other hand, the quality of village life is improved as **remittances** are sent home and used to repair housing and buy household goods.

This case study illustrates how immigrants so often become 'pawns' in the policies and politics of governments. Put bluntly, immigrants generally are warmly welcomed by governments when and where it is felt that there is a need to be met. When the niche is filled or disappears, the welcome quickly chills. Rarely are they welcomed for humanitarian reasons. The cold view of so many governments is that immigrants are fine so long as they are net producers rather than net consumers, or so long as they do not trigger too much change in society.

The help of Carola Suchatzki in compiling this case study is gratefully acknowledged.

15 Question

Draw a simple annotated graph to show the changing attitudes in Germany to Vietnamese immigrants.

Guidance

- Make the *x* (horizontal) axis of your graph your time line.
- The *y* (vertical axis) should be used for the dependent variable, i.e. the changes in attitude. The scale might be labelled from 'cool' to 'warm' or from 'hostile' to 'welcoming'.
- Plot the changes in attitude.
- Mark when the attitude changes with a vertical arrow above the plot.
- Above each arrow, write notes explaining the cause of the shift in attitude. It will look neater if you box these notes.

Using case studies

Consumption

The two issues that recur under the heading 'consumption' are overpopulation and underpopulation. It might be said that overpopulation (in the sense of a mismatch between the number of mouths and available food) is more of a global problem. For this reason, perhaps we should look to the United Nations and its agencies to set the agenda and policies. On the other hand, underpopulation is a problem for a relatively small number of countries. Therefore, policy is much more a matter of what the governments of those countries decide to do, alongside tackling the universal consumption issues such as the motor vehicle (*Case study 13*) and obesity (*Case study 12*). Underpopulation is said to exist where resources and development could support a larger population without any lowering of living standards (e.g. Canada and New Zealand), or where a population is too small to develop its resources effectively (e.g. pioneer and wilderness regions). However, recent events are beginning to define a new underpopulation scenario.

Case study 20 A STAGE 5 ISSUE

A potential new form of underpopulation?

Countries reaching what appears to be Stage 5 in the demographic transition model (DTM) are being confronted by underpopulation. The new scenario is well illustrated by two countries on opposite sides of the world — Japan and Italy.

Japan

Population change in Japan has now passed below the replacement level, and the birth rate is about to pass beneath the death rate (Figure 41). Once the latter happens, the total population will begin to fall from its present peak of 127 million. The combination of a shrinking population and an increasingly elderly one will have a growing impact on everything from demand for goods and services to public finances and the structure of the labour force. For example, within 5 years, Japanese universities expect to have more available places than applicants. The effects on the workforce could be dramatic. The number of people in the economically active age range has been falling since 1995. As the working cohort shrinks to something less than half the population by 2050, so the number of elderly people will rise to account for one-third.

From the point of view of the economy, Japan is fast becoming underpopulated. It is running out of people to keep the economy turning over at its present level. During the next 10 years, the labour shortfall is expected to rise to 5 million.

Japan is also going to become underpopulated in terms of the vast investment that has been made in the past to provide a range of services for a younger population. Schools, colleges and universities are one such category of service. More and more of the infrastructure is becoming surplus to requirements. On the other hand, 'delivering'

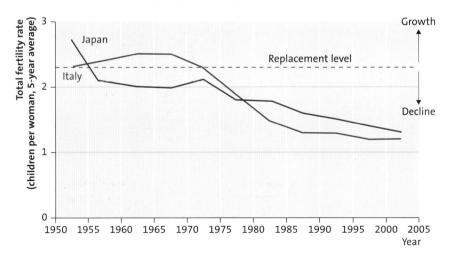

Figure 41 Japan and Italy: changing total fertility rates

Contemporary Case Studies

all that is required by its 'greying' population cannot be guaranteed when the economy is spluttering through lack of workers. So, perhaps we begin to see a paradox. Japan is underpopulated in terms of the economy and young people, but it is clearly becoming overpopulated by the elderly.

The Japanese government has yet to formulate a policy to cope with this emerging scenario. There are a number of options being considered:

- Meet the labour shortfall by attracting immigrant workers — not a popular idea with many Japanese people.
- Make it easier for women and the elderly to take productive jobs.
- Raise pension contributions and cut pensions.

Italy

The demographic situation in Italy is very similar to that of Japan. Italy has one of the lowest levels of fertility in the world. The total fertility rate is 1.23 children per woman, which is far below the replacement level of 2.3 children (Figure 42). Demographers calculate that by 2050 the current population of 57 million could have dwindled to 41 million. If so, towns and cities could be left with thousands of unwanted apartments, some schools may well be half empty and large tracts of the countryside could be depopulated. By 2050, there could be one pensioner for every one productive worker. It raises the interesting question: how is Italy going to support so many pensioners?

It is perhaps worth asking the question: why does Italy not go for the UK solution of mildly encouraging immigration? After all, there are thousands of Albanians just across the water simply itching for a better life. What do you think of this possible solution?

Figure 42 *Total fertility rates and population prospects, 2000*

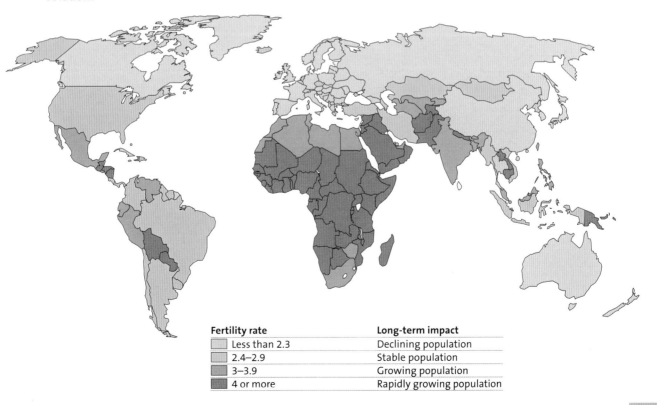

Fertility rate	Long-term impact
Less than 2.3	Declining population
2.4–2.9	Stable population
3–3.9	Growing population
4 or more	Rapidly growing population

Europe

As shown by *Case study 22* and Figure 42, this is not just Italy's problem — it is Europe's too. As yet, none of the countries sharing the problem has policies in place to confront the situation. In 2003, Italy introduced the 'baby bonus' of €1000 for mothers who have a second child. This has now been extended to each first child. Some local authorities have gone further and launched 'baby-making' schemes of their own. For example, in one locality couples are being offered €10 000 for every baby born. However, it is unlikely that incentives such as these can really change the population 'landscape'. At the moment, radical policies simply do not exist either here or elsewhere in Stage 5 countries. As one expert has put it: 'Europeans will probably not wake up to the problems until they are all in their wheelchairs and then suddenly realise there is no one to push!' Is this the endgame to this 'new' form of underpopulation?

Rather than being about specific policies, this case study clearly shows that governments are often slow to respond to situations that are clearly going to happen. What are the reasons — politics, complacency or something else?

16

Using case studies

Question

With the use of examples, explain why the relationship between population and resources is so important.

Guidance

First, explain briefly the threefold relationship: underpopulation, optimum population and overpopulation.

The root of a satisfactory answer lies in identifying the consequences of the unsustainable situation when population exceeds the carrying capacity. Consequences in the natural environment include:

- soil erosion
- deforestation
- air and water pollution

Consequences in the human environment include:

- reduced per capita food supply, malnutrition

- substandard housing
- underemployment
- overstretched services
- political and social unrest
- impaired quality of life

Many of these symptoms are supported by the example of Ethiopia (*Case study 2*).

The discussion might move on to make the point that most of these negative outcomes gradually disappear as population is slowly brought into balance with resource availability and consumption. Use might be made of Chile (*Case study 11*) as an example of a country that has the opportunity of doing this by simply 'tweaking' its distribution of population.

Global issues

This part of the book concludes by looking at five topical global issues, each of which is related to one of the first four key ideas. The threats that they pose mean that these issues represent a serious challenge for the governments of the world. In this day and age of the global village, no country is immune — no country can afford to shrink from playing its part in dealing with the issues. They all have a link with the fifth key component — policy. Every government needs to take its share of the remedial action, as do other major players in global events, particularly the transnational corporations (TNCs).

SIX BILLION AND RISING

Can the world cope?

In 1999, global population passed the 6 billion mark. Forty years ago, there was half that number of people, and in 1900 a quarter. Population has been increasing, largely as a result of mortality control. This came about as a result of improvements in:

- medicine and treatment of disease
- access to healthcare
- sanitation and personal hygiene
- diet and food supply

	% of world population		
	1950	**2000**	**2050**
MEDCs	32	20	13
LEDCs	68	80	87
Africa	9	12	20
Asia	55	61	59
Europe	22	12	7
South America and Caribbean	6	9	9
North America	7	5	4
Oceania	0.5	0.5	0.5

It looks as if the population growth rate peaked in the 1960s at 2.0% per annum; it now stands at 1.3%. The rate may be falling, but because it is effectively a compound rate, large increases in population continue to occur each year. Forecasts indicate that even if the current trend in the growth rate continues, global population could reach 9 billion by 2050. Putting a firm brake on future population growth is going to require an equally effective programme of fertility control. This would require:

Table 6
The changing distribution of the global population, 1950–2050

- better availability and greater use of contraception
- more effective education that emphasises the merits of smaller families and the responsibility of everyone to keep fertility in check
- improved opportunities and choices for women

The question — 'can the world cope with so many people?' — is already a burning one, and will become even more so. Increasing poverty, malnutrition and starvation are indicators that even a population of 6 billion may be unsustainable. Those signs are most evident in the LEDCs. Of the 4.8 billion people in LEDCs, nearly three-fifths lack basic sanitation, about a third have no access to clean water, a quarter do not have adequate housing and a fifth have no access to modern health services. The daily calorie intake of a significant percentage is below that required to complete a day's manual work. However, the immediate problem is that at present something approaching 90% of babies born today will be raised in LEDCs. Table 6 shows a major shift of the world's demographic centre of gravity towards the LEDCs and to Africa in particular.

A population of 6 billion is made unsustainable largely by a fundamental mismatch between the distribution of population and population growth on the one hand, and the distribution of food production and economic growth on the other. The chances of ever harmonising these two distributions are very slim. As a consequence, the prospects are bleak, particularly if you happen to live south of the global North–South divide. Set to become even more stark are the two global geographies — one of plenty, the other of hunger — the former enjoyed by a privileged minority, the other suffered by a deprived majority of the world's population.

The next case study may seem a little perverse in that the previous one has just warned us of the need to check fertility in order to curb the growth in global population. However, you should be saying to yourself: 'are there not an increasing number of countries reaching Stage 4 of the DTM? Surely, this is only achieved if fertility falls to a permanently low pitch? Is this not good news? What can possibly be the problem in such countries?' The following case study moves on from *Case study 20*. Rather than dwelling on the problems and need for new policies, it looks at some of the possible benefits of a 'silvered' society.

Case study 22 'GREYING' POPULATIONS

A sustainable new scenario?

The UK is one of 61 countries in which not enough babies are being born to replace their populations. In 24 European countries, and in other MEDCs elsewhere, fertility is now so low that populations are set to decline (Table 7). As should now be well understood, decline leads inevitably to an ageing or 'greying' society. The age–sex pyramid is progressively undercut at its base and the upward taper becomes much blunter. The most obvious outcome of this is a shift in dependency, whereby the elderly outnumber the young. This in turn changes the nature of the demand for social services. Crudely put, demand shifts from crèches and schools to sheltered accommodation and hospital beds.

The ageing society is usually portrayed as a nightmare scenario of hospital beds and care homes packed with frail elderly, of mass poverty among growing ranks of pensioners and of fewer people to provide much-needed care and support. These things might happen and there is no denying that they represent the downside of an ageing society. However, they could be counteracted by making some socio-economic adjustments, including:

- discouraging early retirement
- raising the pension age to at least 65 (perhaps 67 or 70) for both men and women
- lowering the high turnover in labour caused by early retirement and the recruitment of younger employees
- requiring that workers do more to set up their own pension funds, rather than relying on an increasingly inadequate state pension
- changing the mindset of employers to recognise the value of experience in the labour force

Actions along these lines would undoubtedly maintain a larger labour force. They would also improve a country's ability to support growing numbers of elderly people.

Even without these changes, there is much to celebrate in today's greying societies — increasing life expectancy, longer retirement, more choice. A profile of 'ancient Britain' in 2000 includes the following facts:

Table 7 *Getting smaller: Europe's projected population decline, 2000–50*

Country	Population, 2000 (thousands)	Population, 2050 (thousands)
Austria	8 142	7 094
Belgium	10 301	8 918
Bulgaria	7 707	5 673
Croatia	4 297	3 673
Czech Rep.	10 197	7 829
Denmark	5 375	4 793
Finland	5 187	4 898
Germany	81 895	73 303
Greece	10 559	8 233
Hungary	10 029	7 488
Italy	57 513	41 197
Lithuania	3 582	2 297
Netherlands	16 078	14 156
Poland	38 646	36 256
Portugal	10 084	8 137
Russian Fed.	143 300	121 256
Serbia-Montenegro	10 701	10 548
Slovakia	5 400	4 836
Spain	40 586	30 226
Sweden	8 924	8 661
Switzerland	7 319	6 745
Ukraine	49 927	39 302
UK	59 734	56 667

- There were 10.7 million people aged 50 or over.
- 5.4 million women were aged 65 or over compared with 3.9 million men.
- A man aged 60 could expect to live for another 19 years and a woman another 23 years.
- Of men aged over 65 and of women over 60, 8% were in employment.
- Since 1980, the numbers of over-50s in work had fallen by 20%, and the numbers of over-60s by 30%.
- 70% of pensioners depended on state benefits for over 50% of their income.
- Of people aged over 75, 29% of men were widowed, compared with 61% of women.

A combination of good health, paid-up mortgages and 'empty nests' means that many people in their 60s and 70s are taking up new interests, going back to the classroom and fuelling a boom in the leisure market. A significant growth of overseas 'ski' ('spending the kids' inheritance') travel is perhaps spreading benefits to those LEDCs currently viewed as attractive tourist destinations.

It is anticipated that popular culture will change. When over half the population is over 50, it will no longer be dominated by the obsession with youth. Society's institutions will have to adapt to the interests of older people. Most crimes are committed by the young, and the elderly are far more law-abiding, so the crime rate is expected steadily to fall. There will be less need for police and prisons. Politics will become more stable and less prone to knee-jerk reactions and seesaws in policy. The 'grey' lobby will grow ever more powerful and difficult to ignore. Elderly people have more experience; they have seen it all before; they are more conservative. Elderly people are becoming more participative and proactive in politics, voluntary work and community associations.

The changes of a 'greying' society are likely to bring parents and children closer together, changing the relationship from one of dependency to one of equality. Longer life expectancy means that parent and child now enjoy a longer adult relationship free of the reciprocal dependency responsibilities of either child rearing or caring for the elderly in their last years.

At present, the scenario just painted applies only to a small, but growing, number of MEDCs. One wonders how long will it be before it becomes a truly global issue: or might it be that in many of today's LEDCs the AIDS pandemic will delay it indefinitely?

17
Using case studies

Question

Discuss the view that the challenge of a 'greying' population is now a global issue.

Guidance

Figure 43 demonstrates the spider-diagram technique that can be usefully applied in planning essays and answers to examination questions.

Having 'thought-bombed' (brainstormed) the key points you think need to be made (Figure 43), it is then necessary to order them into a coherent sequence. Alternatively, you might rank them on the basis of their importance to your discussion.

Try to number the boxes in Figure 43 and write an answer. Of course, you might wish to add more points to the diagram before you start writing.

The DTM suggests that most, if not all, countries will eventually reach at least Stage 4. Europe and North America today — where tomorrow? If true, certainly this will make 'greying' a global issue.

'Greying' societies mean MEDCs are less able to help LEDCs. This helps make the issue a global one.

'Greying' countries suffer increasing labour shortages. Who works the economy and who provides services, especially those for the elderly?

Immigration is a short-term solution to labour shortages. Presumably immigrants will come from current LEDCs. This helps make the issue a global one.

What happens when the labour shortage becomes a global one? World governments will have to promote pro-natalist policies largely by means of incentives.

'Greying' populations hold out the promise of a global future that is sustainable so far as the environment and resources are concerned.

POSSIBLE CONCLUSION

The issue is becoming a global one. It looks as if the world may be sitting on a demographic time bomb.

Figure 43
A spider diagram to use as an answer plan

Case studies 21 and 22 paint what might seem to be a conflicting picture of the world — a youthful and highly reproductive population versus an ageing and declining population. Each situation is true, but only for a portion of the globe. Their side-by-side existence is one of a number of factors underlying the next of our global issues — the migration explosion.

Case study 23 — A WORLD OF MIGRANTS

The right to roam?

Globally, migration has doubled over the past two decades and seems likely to continue accelerating, affecting every country in the world. The number of migrants worldwide increased from 84 million in 1975 to 175 million in 2000 (Figure 43). Current predictions say that it will rise to 230 million by 2050.

About 23 million people emigrate from LEDCs to MEDCs each year. These migrants now account for two-thirds of the population growth in the North. The migration dream is being encouraged in a variety of ways. Imported television programmes and the advertising of imported luxury goods, plus talk of high wages, persuade many LEDC citizens that a much more comfortable lifestyle is to be found in Europe, the USA and Australasia. However, the main driving force behind the South–North movement of people comes from the reciprocal demographic situations illustrated by *Case studies 21* and *22*. On the one hand, the rapid population growth in the South increases the pressure to emigrate to the North in search of work. At the same time, labour, particularly unskilled labour, is in increasingly short supply in the 'greying' populations of the North. In short, a one-way labour migration is helping to solve both problems.

Three other factors have played a part in raising levels of migration:

■ Some political barriers have been removed, for example in the break-up of the Soviet Union and China's relaxation of travel restrictions.
■ Cheaper long-distance transport facilitates international migration.
■ The communications revolution (television, the internet, mobile phones) has resulted in a much greater awareness of opportunities overseas.

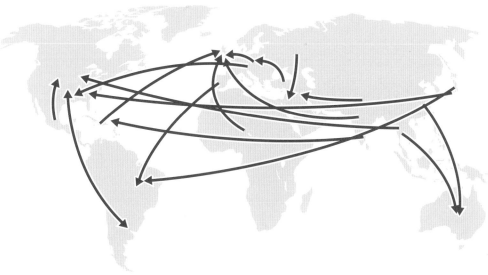

Figure 44 *The world on the move*

However, not all of today's international migrants are searching for work. Since the end of the Second World War in 1945, there have been over 150 other wars, mostly in LEDCs. These wars, together with persecution policies in some countries (e.g. Bosnia, Serbia, Kosovo and Iraq), have generated great streams of refugees. People have abandoned their homes in fear. It is estimated that there are some 40 million refugees in the world today.

For centuries, Europe was the continent of emigration. That has now changed. Asia has become the number one source of emigrants, the majority coming from India, Bangladesh, Pakistan, China, Vietnam and the Philippines. At the same time, Europe has become the immigrant honeypot. Here, 56 million people (nearly 8% per cent of the population) are living outside their country of birth. Although the absolute numbers are smaller in North America and Oceania (Australasia), the percentage figures are higher, at 13% and 19% respectively (Figure 44). In contrast, the South is relatively unaffected by migrants. They account for only about 1% of the population of Asia and Latin America, and 2% of Africa.

The present global migration from LEDCs to MEDCs can be seen, not just as an acceleration, but also as a continuation of migration flows that have waxed and waned over the centuries; flows that have changed the face of the world. Figure 44 identifies some of the more trodden migration pathways of the last 100 years.

Government policies can have a major impact on international migration. They can either make it happen or prevent it. The latter is more common, as more and more countries are either closing their borders or severely restricting in-flow. Twenty-five years ago, only 6% of countries had policies to curb immigration; now 40% of countries do. There are only five countries in the world today that have official policies to encourage permanent in-migration. They are the USA, Canada, Australia, New Zealand and Israel. Recently, the UK has been added to that list, since it has said that it wants 150 000 immigrants a year.

One of the directors of the International Organisation for Migration said recently: 'Migration is a reality; the issue is how to make it work to benefit both migrants and the receiving populations.' Unfortunately, there are other issues that also need to be sorted out, such as bogus asylum seeking (*Case study 28*), illegal trafficking in people and tighter border controls that threaten the human right to freedom of movement.

The rising tide of global migration is mainly the outcome of the widening gap between the rich and poor nations of the world. Differences in living standards, welfare and opportunities for betterment create a powerful alignment of pull and push factors. They are the very stuff on which migration feeds. The next case study focuses on an issue that is largely about consumption, but one that is also closely related to population growth and distribution.

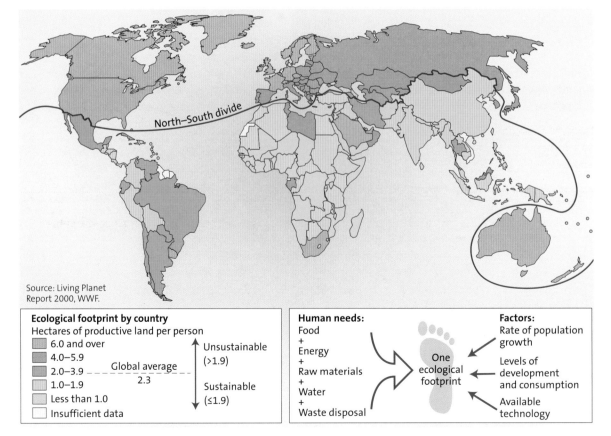

Case study 24 THE ECOLOGICAL FOOTPRINT

Can we stop it getting deeper?

The ecological footprint (EF) is a measure of a population's consumption of renewable natural resources. It is usually calculated in terms of the total area of productive land and sea required per person to meet their food, energy, raw material, water and waste disposal needs. The 'depth' of the footprint is conditioned by three key factors:

- the rate of population growth
- the levels of development and consumption
- the nature of available technology

The EF of a population can be calculated at a range of spatial scales from the global and regional to the national and local.

Figure 45 The global ecological footprint

The Earth has about 11.4 billion hectares of productive land and sea space — about a quarter of its surface area. Divided between a global population of 6 billion, this

Source: Living Planet Report 2000, WWF.

North–South divide

Ecological footprint by country
Hectares of productive land per person

- 6.0 and over
- 4.0–5.9
- 2.0–3.9
- 1.0–1.9
- Less than 1.0
- Insufficient data

Global average 2.3

Unsustainable (>1.9)

Sustainable (≤1.9)

Human needs:
Food
+
Energy
+
Raw materials
+
Water
+
Waste disposal

One ecological footprint

Factors:
Rate of population growth

Levels of development and consumption

Available technology

equates to 1.9 ha per person. According to a recent report by the World Wide Fund for Nature (WWF), while the EF of the average African and Asian consumer was less than 1.4 ha per person in 2000, the average western European's footprint was 5.0 ha and the average North American's 9.6 ha. The mean EF of the global consumer was 2.3 ha per person. This is 20% above the Earth's biological capacity of 1.9 ha. This means that the human race now exceeds the world's ability to sustain its consumption of renewable resources. Currently, the human race is only able to maintain this 'global overdraft' on a temporary basis by eating into the Earth's capital stocks of forest, fish and fertile soils. We are also dumping our excess carbon dioxide into the atmosphere. Neither of these two activities is sustainable in the long term. The only sustainable solution is to live within the biological capacity of the Earth. If we do not, the outlook can only be one of catastrophe.

The current trends show that the human race is moving away from achieving the minimum requirement for sustainability, not towards it. The global EF grew from about 70% of the Earth's biological capacity in 1960 to about 120% of its capacity in 2000. Projections suggest that the EF is likely to grow to around 200% of the Earth's biological capacity by the year 2050. However, it is very unlikely that the Earth would be able to run an ecological overdraft for another 50 years without some serious backlashes in terms of human welfare, economic development and the environment.

If the human race is to follow a sustainable development pathway, it will involve making changes in four fundamental ways. We must:
- improve the efficiency with which resources are used in the production of goods and services
- reduce the 'widening gap' between high- and low-income countries, particularly by reducing the high consumption levels of the former
- protect, manage and restore natural ecosystems and so maintain (or even enhance) biological diversity and biological productivity
- control population growth through the promotion of contraception, education and healthcare

No nation can go it alone. Achieving a sustainable future requires everyone to play their part in helping to reduce the global ecological footprint. What are *you* prepared to do about it?

18

Using case studies

Question 1
Which of the four changes bulleted in *Case study 24* is most crucial in finding a sustainable development pathway?

Question 2
Which one of the four global issues outlined in *Case studies 21, 22, 23* and *24* do you think presents the greatest global threat?

Guidance
These two questions are for you to debate with other members of your group. Organise yourselves so that one or two students make a presentation in support of one of the four changes or issues. If you wish to convince others, you will need to marshal supporting examples. Perhaps your teacher might act as judge and jury!

Illustrating the links

In parts 1 and 2 of this book we looked at population in terms of four components — distribution, change, migration and consumption — held together by a single theme of policy (Figure 1). In reality, these components are interlinked. For example, the distribution of population is affected by the patterns of population change and migration. In turn, population change and migration impact upon the pattern of consumption, if only because increasing the concentration of people in particular locations will inevitably raise the levels of consumption. The purpose of this part of the book is to explore and illustrate some of these links.

Migration and distribution

The links here are fairly evident. Migration might be expected to change the distribution of population, lowering it in source areas and raising it in destination areas. However, in the case of international migration, what, if any, are its impacts on the population distribution pattern of the receiving country? Since much migration typically involves people in the reproductive age range, what, if any, is the impact on fertility and therefore rates of population change? Are lower birth rates plus reduced growth (perhaps even decline) at the source and raised birth rates plus increased growth at the destination the most likely scenarios?

Case study 25 NUMBERS AND ETHNICITY CHANGE, BUT LITTLE ELSE

The case of Australia

'Dead heart' and 'Polo mint' are two frequently used descriptions of the pattern of population distribution in Australia. Ever since the early colonial development of this country, most of the settlement has taken place in coastal regions, particularly along the eastern and southwestern seaboards (Figure 46).

Although Australia had, and still has, an aboriginal population (currently estimated as around 0.5 million), much of the peopling of this vast country has been through immigration. Up until the end of its 'whites only' immigration policy and assisted passage schemes in the 1960s, most new settlers came from the UK. This is still reflected in the fact that roughly a quarter of all of today's 'overseas-born' Australians started their lives in the UK. While the figure of around a quarter also applies to the present inward flow of migrants, it conceals the fact that there has been a fundamental shift in the 'sourcing'

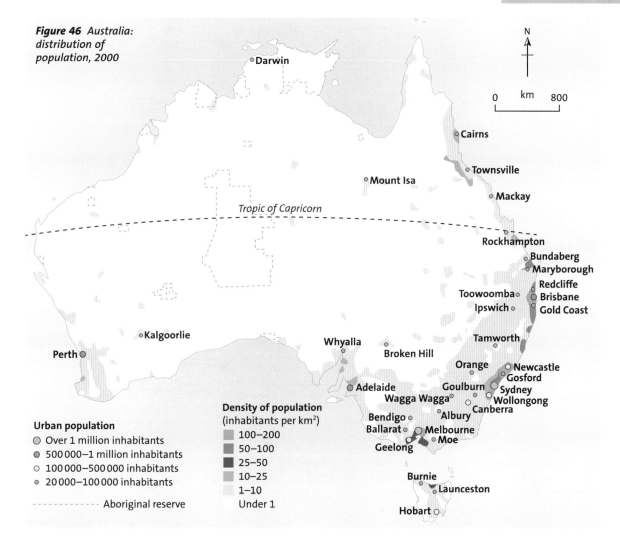

Figure 46 Australia: distribution of population, 2000

Darwin

Cairns

Townsville

Mount Isa

Mackay

Tropic of Capricorn

Rockhampton
Bundaberg
Maryborough
Redcliffe
Toowoomba
Brisbane
Ipswich
Gold Coast

Kalgoorlie

Whyalla

Tamworth

Perth

Broken Hill

Orange
Newcastle
Gosford
Adelaide
Goulburn
Sydney
Wagga Wagga
Wollongong
Canberra
Bendigo
Albury
Ballarat
Melbourne
Geelong
Moe

Burnie
Launceston

Hobart

Urban population
- ◉ Over 1 million inhabitants
- ● 500 000–1 million inhabitants
- ○ 100 000–500 000 inhabitants
- ∘ 20 000–100 000 inhabitants

- - - - - Aboriginal reserve

Density of population
(inhabitants per km²)
- 100–200
- 50–100
- 25–50
- 10–25
- 1–10
- Under 1

N

0   km   800

of settlers. Today, nearly half come from outside Europe, and half of these are drawn from various parts of Asia (Figure 47).

The data in Tables 8–11 show how the national origins of Australian immigrants have changed over four decades, most notably:

- the substantial decline in numbers from the UK
- the rise of New Zealand as the number one source
- the overall shift of emphasis from Europe to Asia

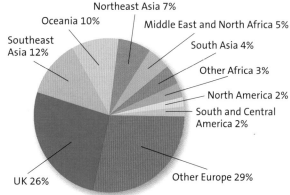

Northeast Asia 7%
Oceania 10%
Middle East and North Africa 5%
Southeast Asia 12%
South Asia 4%
Other Africa 3%
North America 2%
South and Central America 2%
UK 26%
Other Europe 29%

Figure 47 Australia: overseas-born population by region, 1999

Despite the steady inward movement of migrants over the last 30 years (averaging around 70 000 each year), natural increase has been the main contributor to the growth in Australia's population. Natural increase has not varied greatly in absolute terms, but

Birthplace	Arrivals (thousands)	%
UK and Ireland	361.6	50.4
Greece	67.0	9.3
Italy	63.2	8.8
Yugoslavia	38.5	5.4
Germany	18.3	2.5
Malta	16.5	2.3

Table 8 *Australia: birthplaces of settler arrivals, 1964–68*

Birthplace	Arrivals (thousands)	%
UK and Ireland	142.6	35.8
New Zealand	22.6	5.7
Lebanon	21.7	5.4
Yugoslavia	15.5	3.9
Greece	10.5	2.6
USA	10.1	2.5

Table 9 *Australia: birthplaces of settler arrivals, 1974–78*

Birthplace	Arrivals (thousands)	%
UK and Ireland	91.9	18.5
New Zealand	62.6	12.6
Vietnam	37.8	7.6
Philippines	27.0	5.4
South Africa	14.7	3.0
Poland	7.7	1.5

Table 10 *Australia: birthplaces of settler arrivals, 1984–88*

Birthplace	Arrivals (thousands)	%
New Zealand	58.3	13.9
UK and Ireland	53.6	12.8
China	29.8	7.1
Vietnam	19.4	4.6
Hong Kong	19.0	4.5
Philippines	17.1	4.1

Table 11 *Australia: birthplaces of settler arrivals, 1994–98*

in contrast net migration has fluctuated markedly (Figure 48). There are only 3 years in which immigration exceeded natural increase as the main component of population growth. What is also interesting is that the shift in the sourcing of settlers from Europe to Asia does not appear, as yet, to have had much impact on the birth rate, which continues to fall from the high of 23 per 1000 in the 1950s. It seems that the MEDC perception that the smaller the family, the better the quality of life has caught on among the new migrants. The birth rate currently stands at 14 per 1000.

Figure 48 *Australia: components of population growth, 1972–99*

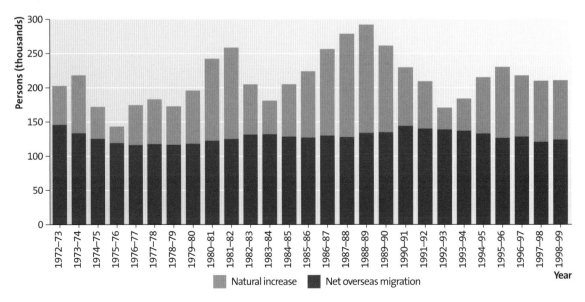

Recently, in an attempt to encourage settlement in the interior, the Australian government has introduced a scheme whereby virtually anyone is allowed to enter the country provided they spend their first 10 years living in the outback. It remains to be seen what impact this might have on the distribution of population.

The bottom line to this case study is that, while the population of Australia has grown considerably (from 8 million to nearly 20 million in 50 years) and the origins of immigrants have changed, there has been no substantial alteration in the traditional pattern of population distribution. The physical hostility of the country's desert interior and the momentum of settlement along the coast have resulted in Australia retaining its 'dead heart'.

Distribution and consumption

The link between population distribution and consumption might seem to be simple. You could argue that the more people there are in an area (the higher the population density), the greater will be the consumption of resources. However, the relationship is complicated by a number of factors. Perhaps the most important of these is the level of development, in that the more advanced and the richer a country is, the higher is the per capita consumption of resources (food, minerals, energy etc.). The next case study illustrates this point by putting the spotlight on the world's demographic heavyweights. It also indicates that current rates of population growth may have a direct bearing on consumption and future levels of human development. Three issues raised in earlier case studies (*Case studies 22, 23* and *24*) are also drawn into the discussion.

THE WORLD'S MOST POPULATED COUNTRIES

Case study 26

The global heavyweights

In August 1999, the world's population passed the 6 billion mark. Less than a year later, India joined China as the second country with a population over 1 billion. Projections indicate that in 50 years' time, India will surpass China as the world's most populous nation, when it is expected to have 1.5 billion inhabitants. The basic reason for the change in ranking is that while China has a family planning programme (see *Case study 17*), India's attempt to put one in place has been a colossal failure; in India today, a baby is born every 2 seconds! Its population is increasing by 2.5% a year, compared with 0.9% in China (Table 12). The present mean population density figure is 307 persons per km^2 compared with 244 for the UK and 133 for China. However, none of these densities comes anywhere near the 953 persons per km^2 recorded in Bangladesh.

Table 12 gives information about the current ten most populated countries in the world (the UK ranks number 20). The size of the population of China and India is emphasised by the huge gap between them and the third ranking country, the USA. Together the ten countries account for just under 40% of the world's land area and around 60% of its population. In terms of distribution, seven of the countries are

Table 12 The ten most populated countries in the world

Rank	Country	Population, 2000 (millions)	Density, 2000 (people/ km²)	Mean annual population change, 1995–2000 (%)	HDI, 2000	Ecological footprint, 2000*
1	China	1 262	133	0.9	0.726	1.54
2	India	1 014	307	2.5	0.577	0.77
3	USA	276	29	1.0	0.937	9.70
4	Indonesia	225	111	1.8	0.684	1.13
5	Brazil	173	20	0.8	0.757	2.38
6	Russia	146	9	0.0	0.781	4.55
7	Pakistan	142	176	2.8	0.499	0.64
8	Bangladesh	129	953	1.0	0.478	0.53
9	Japan	127	336	0.3	0.932	4.77
10	Nigeria	123	56	1.2	0.462	1.33

*See Figure 45

located in Asia (the whole Indian subcontinent is included), two in the Americas and one in Africa.

With respect to consumption, it is significant that three of the countries (the USA, Russia and Japan) are MEDCs with typically high levels of resource consumption, which is reflected in their high ecological footprint values (see Figure 45). In Russia and Japan, low rates of population increase may help to ease the situation. However, the much higher rate in the USA, with one of the heaviest ecological footprints, is worrying, as are the even higher rates in India, Pakistan and Indonesia. Increased population may not necessarily mean ecological crisis, but it will not ease the pressure on already stressed ecosystems. It is bound to create competition for, if not conflict over, scarce resources such as fresh water, farmland, minerals and timber.

There is double trouble in store. The rich nations of the world, with their high consumption levels, have a record of polluting and destabilising the world far more than the poor. This is unlikely to change. Equally, land degradation and desertification are real and growing problems in LEDCs, many of which are too poor to be able to cope with the challenges. Children born into at least half these countries can look forward to a rough ride in life. Unless there is major change in the world, most of these children will live in cities and in poverty. There will be little or no narrowing of the range in human development index (HDI) values. There will also be widespread food shortages, as well as serious sanitation and health problems.

Migration and consumption

As can be seen from Table 12, Indonesia has the fourth largest population in the world. However, there is nothing spectacular about its population density figure of 111 persons per km². In reality, this mean figure conceals the fact that some of Indonesia's many islands suffer from acute overpopulation, while others are distinctly underpopulated. The next case study looks at an attempt that was made to even out these disparities by encouraging migration.

INDONESIA'S TRANSMIGRATION POLICY

An appraisal

Transmigration is not a new policy. It was initiated during Dutch colonial rule in the early twentieth century and taken over by the Indonesian government after independence in 1949. The three main goals of transmigration were:

- to move millions of Indonesians from the very overpopulated inner islands (Java, Bali, Madura) to the outer, less-densely populated islands in order to achieve a better distribution of population (Figure 49)
- to alleviate poverty by providing land and other opportunities so that landless settlers could generate income
- to exploit more effectively the 'potential' of the underpopulated outer islands

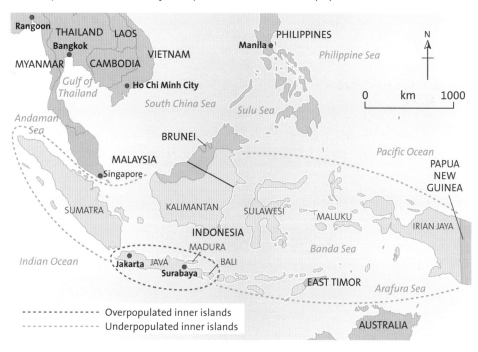

Figure 49
Map of Indonesia

------------- Overpopulated inner islands
------------- Underpopulated inner islands

Additional targets gained importance during General Suharto's regime, which held power for 30 years from 1966, namely:

- regional development
- nation building
- national security

Under this regime, transmigration increased dramatically and large numbers of people were resettled (Table 13), mainly to Kalimantan, Sumatra, Sulawesi, Maluku and West Papua (Irian Jaya). The massive financial support from the World Bank, the Asian

Table 13
Transmigration data, 1950–2001

	1950–69	1969–74	1974–79	1979–84	1984–89	1989–94	1994–99	1999–2001
Target (families)	—	38 700	250 000	500 000	750 000	550 000	600 000	16 250
Families moved	100 000	36 483	118 000	535 000	230 000	No data	300 000	4 400
Number of people	500 000	174 000	544 000	2 469 600	1 061 700	No data	1 500 000	22 000

Development Bank and bilateral donors helped to boost the programme in the 1980s. However, this expansion of the programme began to alert environmental and human rights critics both inside and outside Indonesia. They revealed that transmigration was in effect a development fraud and an environmental disaster rather than a humanitarian programme aimed at achieving a more equitable distribution of population.

Critics of the transmigration policy suggest that:

- it has had a devastating impact on the rainforest — Indonesia's outer islands contain some 10% of the world's remaining stock
- it was politically inspired and used to control the indigenous population of the outer islands (e.g. in Irian Jaya, East Timor, Kalimantan)
- it has violated traditional land rights and was designed to force the assimilation of indigenous people and forest dwellers
- it has been an economic disaster, increasing Indonesia's national debt and, in some years, swallowing 30–40% of the economic development budget of the outer islands
- poverty has not been alleviated but redistributed — most transmigrants are worse off because of inadequate planning and site preparation, poor access to markets, and neglect of soil and water properties indispensable to an agricultural economy
- it has hardly made any dent in the population pressure in Java
- it has created environmental havoc on the outer islands

Various internal and external factors have changed the transmigration policies over recent years. In the 1990s external support dried up as the funding organisations became aware of the programme's failings. However, resettlement figures remained high during the first half of the decade (see Table 13). At the same time, foreign financial assistance switched to a new strategy to support 'second-stage' transmigration, i.e. rehabilitation of the existing resettlement projects. A severe financial crisis ('a dose of Asian flu') hit Indonesia in 1997. The ensuing struggle to rebuild the economy and reform a corrupt political system has resulted in major changes in the political and economic landscape. In turn, these have substantially influenced the transmigration programme over the past few years. The picture is complex and dynamic; it is both reassuring and alarming. On the positive side, the official transmigration programme, as implemented during the Suharto years, appears to have been quietly dropped by the current government. The new political openness means that the coercive elements of transmigration have been neutralised. However, there is a real danger that transmigration in a new guise may take over where the old programme left off. The central government and the newly empowered local governments are relying on natural resource exploitation — logging, mining, industrial timber and pulpwood plantations, oil palm, and industrial shrimp farming — to generate revenue. This large-scale commercial exploitation aimed at export markets is being actively encouraged by Indonesia's international creditors, led by the International Monetary Fund and the World Bank. If this continues, the demand for labour in areas of low population will increase, which could fuel a new transmigration boom.

This case study shows close links between policy and population distribution and consumption. What started as a genuine attempt to overcome the twin problems of over-population and underpopulation was hijacked by a corrupt political regime. Resource exploitation and the subjugation of indigenous peoples became the overriding motives, resulting in immense environmental damage and serious abuse of human rights.

Two overarching case studies

Finally, we take a look at two demographic issues, both of which are highly topical and of immense importance. The first issue concerns us greatly in the UK today, namely, the arrival of large numbers of asylum seekers. The second is AIDS, and the critical need to take every measure possible to halt its spread and devastation, particularly in Africa. Both issues are the product of globalisation. Many asylum seekers are simply part of the global movement of labour. The spread of AIDS is partly the product of increased global travel. Dealing with both issues requires concerted actions by the whole global community. These include:

■ respecting the rights of genuine refugees
■ giving poor countries ready access to the same anti-HIV drugs as rich ones

Both issues impact on all those five aspects of population geography, namely, distribution, change, migration, consumption and related policies.

THE UK AND ASYLUM SEEKERS
Case study 28

A complex and sensitive issue

During the 1990s, apart from one year, the annual international migration balance of the UK showed a net gain (Figure 50). However, when the distinction is drawn between British and non-British citizens, two very different situations emerge. The record for the former has been one of persistent net loss, while that of the latter has been one of considerable net gain. Among the non-British citizens are increasing numbers of asylum seekers.

Figure 50
International migration into and out of the UK, 1991–2001

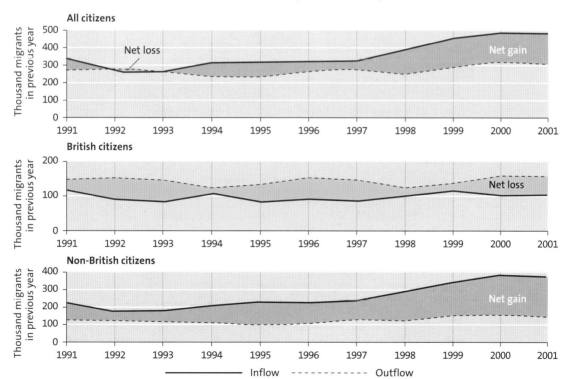

Inflow ––––––– Outflow

Between 1982 and 2001, the number of people seeking asylum in the UK showed a 20-fold increase, from around 4000 to 80 000 per year. The more recent asylum-seekers are part of a broad movement of population towards Europe (particularly to the EU) from Africa, the Middle East and Asia. To a lesser extent, they represent migration within Europe away from the trouble-torn former Yugoslavia and the poverty-ridden countries of the former Soviet bloc. The most worrying aspect for the UK is that it is the most popular asylum-seeker destination. The reasons for this are certainly to do with personal perceptions and possibly a reputation that the UK is a 'soft touch'. However, the arrival of such huge numbers of asylum seekers at our borders has raised a fundamental issue — should all asylum seekers be allowed to enter and settle? In order to understand this issue, it is vital to be clear about:

- the difference between an asylum seeker and a refugee
- the immigrant pathways into a country

An **asylum seeker** is a person who seeks to gain entry to another country by claiming to be a victim of persecution, hardship or some other compelling circumstance. A **refugee**, as defined by the UN, is someone whose reasons for moving are genuinely to do with fear of persecution or death. Figure 51 shows three broad pathways into a country. Particularly important here is the 'conversion' of an asylum seeker into a refugee. Although official statistics are hard to come by, it appears that less than half of all asylum seekers are granted leave to stay in the UK.

Figure 51
Immigrant pathways

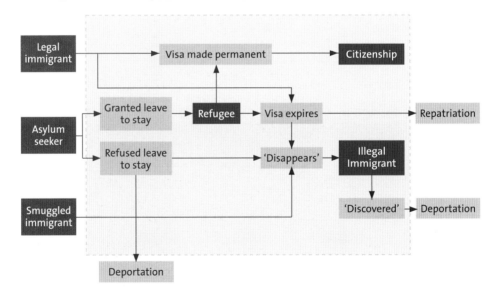

Figure 52 looks at the issue of asylum seeking from three different viewpoints currently held in the UK. Those viewpoints — admit none, admit some and admit all — occupy different points along what might be called the 'immigrant welcome scale'.

The 'all-are-welcome' viewpoint is based on the belief that freedom of movement is a basic human right, as stated in the UN Declaration of Human Rights (1948). Most subscribers to this view would argue that, having entered a country, this 'freedom' should extend to allow migrants to choose where they settle. A minority would accept that for practical reasons (availability of housing, work and services) there has to be some degree of intervention in, or management of, location.

Who to admit?	Why?	Where to locate?
NONE	Reason 1: the 2001 census shows that while the UK is becoming more multicultural, second- and even third-generation immigrants are still among the most deprived in society (*Case study 10*). The figures show that black, Asian and other ethnic minorities are twice as likely to be unemployed, half as likely to own their own home and run double the risk of poor health, compared with white Britons. The census data undermine the belief that first-generation immigrants may suffer, but their subsequent British-born offspring move off the breadline and become better off. Reason 2: in 2003, an all-party group of MPs claimed that the huge numbers of asylum seekers entering Britain were threatening the country and likely to spark social unrest. They said that the situation had already provoked a political backlash, with voters turning to extremist parties in protest against the spiralling numbers who have come to Britain during the last 20 years. If the current rate of entry persists, the country's capacity to cope, in terms of providing housing, work and social services, would be overwhelmed. Better to close the doors now than run the risk of social unrest.	Send back
SOME Refugees	It is clear that not all asylum seekers are genuine refugees (people whose reasons for moving are to do with a real fear of persecution or death). The fact that so many asylum seekers entering the European Union make for the UK raises the suspicion that many of those claiming to be refugees are in fact 'opportunists' looking for work and a better life. Equally, the fact that most come from Iraq, Zimbabwe, Somalia, Afghanistan and China suggests that persecution may be the key push factor. The challenges are to discriminate between refugee and opportunist; to process asylum applications as quickly as possible; to treat all applicants humanely while this is being done; to ensure that those denied leave to stay are 'removed' and do not 'disappear', swelling the number of illegal immigrants.	Concentrate or disperse
Close family	There are powerful humanitarian arguments for admitting close relatives of people already living legally in the UK. The problem is how far to extend the family net. Should it be limited to next of kin — parents and children — or extended to aunts, uncles and cousins? Then there is the delicate issue of marriage. Should a line be drawn between 'arranged' and 'chance' marriages, and between foreigners entering the country as a national's partner, rather than as a legal spouse?	Where appropriate
Key workers	The British government recently set up the Highly Skilled Immigrant Programme to meet shortfalls in the skills market. So far it has attracted 3000 well-qualified foreigners and their dependants. The scheme makes it easier for the brightest foreign students to carry on working in the UK after their courses finish. Work permits are given readily to international business executives. The government has stated that it wants to admit 150 000 key workers a year as part of the programme. This selective-door process is a reversal of the closed-door policy that has prevailed since 1970. In the 1950s and 1960s, when cheap, unskilled labour was in short supply, the government had operated an open-door policy, particularly to immigrants from Commonwealth countries in Africa and the Caribbean.	Where needed
ALL	Many people believe that freedom of movement is a basic human right — it was enshrined in the UN Declaration of Human Rights (1948). The view of civil liberties organisations around the world is that this right is paramount. It should be upheld, even to the point of disregarding any political, economic, social or practical problems. However, there are few governments today actively supporting the right, other than when a government is keen to rid its country of a particular group of people (e.g. pre-war Germany and the Jews) or when governments are concerned about the safety of minority groups in another country (e.g. Muslims in Serbia).	Where they choose

Cold — Welcome scale — Warm

Figure 52 Asylum seekers in the UK: attitudes and issues

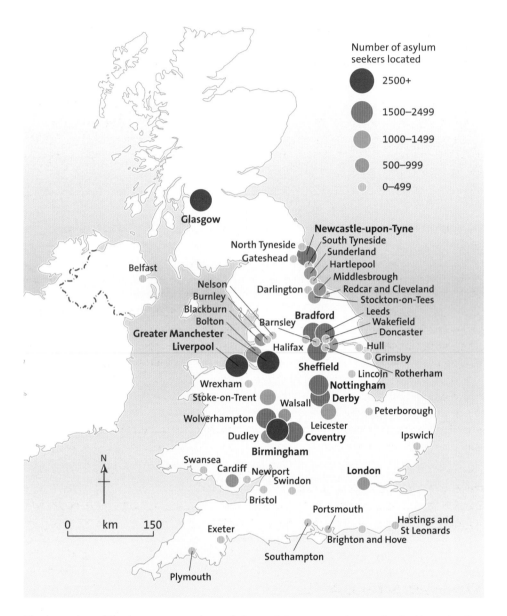

Figure 53 *Where asylum seekers are sent*

The keynotes of the 'some-are-welcome' viewpoint are screening and selection, starting from three different bases:

- discriminate between the genuine refugee and the opportunist (economic refugee)
- admit only close relatives of existing citizens
- admit those who may be expected to bring some 'benefit' or meet a particular demand — specific skills, investment capital, cheap labour

Taking the first bullet point, there are two conflicting views about what to do with the genuine refugees — concentrate or disperse them. Considering the second and third bullet points, there would be general agreement that the question of 'where to locate' should be left to family choice and market forces. The result could be either dispersion or concentration.

Contemporary Case Studies

Arguments for the 'none-are-welcome' viewpoint focus on:

- the likely plight of immigrants — discrimination, persecution, poverty and hardship, inadequate housing, unemployment
- the impact of immigrants — on housing, unemployment, over-stretched services

Each of the viewpoints on the admission of asylum seekers into the UK (Figure 52) can be supported, to varying degrees, by the quotation of official statistics. Each can be made to look persuasive. Who is to say which of these views is right or wrong?

From a geographical viewpoint, perhaps the most interesting aspect of the issue is 'where to locate?'. This relates particularly to two critical stages along the asylum seeker's pathway:

- while the asylum application is being processed
- when permission is granted to stay

Given that most asylum seekers arrive either by ferry at the Channel ports or at London's airports, it is clear that such convergence puts a great strain on reception services, particularly in London and Kent. For this reason, it is now policy to disperse asylum seekers and to hold many of them in designated centres (Figure 53). The policy of dispersal seems to have worked reasonably well. However, the scheme has been suspended in certain areas (e.g. Manchester, Burnley and Bolton) at the request of the police and local councils. This has been justified on the basis of nipping threatened ethnic disturbances in the bud. It is proposals to set up large, sometimes purpose-built, holding or detention centres that have sparked the fiercest local resistance.

The issue of location comes to the forefront again once permission has been granted for the asylum seeker to stay. At this stage, the policy of dispersal has, once more, much to commend it. Surely, this is preferable to a policy of concentration, particularly if the intention is that such incomers should be assimilated into Britain's increasingly multi-ethnic society? Concentration means segregation; segregation means little assimilation.

This issue has many different facets that have not been explored here — facets that are to do with human rights and ethics, party politics and economics. It is important that we should be as fully informed as possible about all aspects of this issue, and not just the geographical dimension that has been brought into focus in this case study.

A GLOBAL DEMOGRAPHIC EMERGENCY

Case study **29**

AIDS in sub-Saharan Africa

Two decades after AIDS was first identified, the pandemic has a deadly grip on the world. In 2004, an estimated 40 million people carried the HIV infection that causes AIDS. Meanwhile, the annual death toll rose from just over 2 million in 1999 to 3 million in 2003. After years of denial that the disease existed, the LEDCs (where most of the affected live) are gradually waking up to the need for action. More political energy is being focused on the challenge. Huge sums of money are being raised (nearly $5 billion in 2003) by a range of donors, from UN agencies and multilateral organisations to individual governments and private groups. This money and the political will are being directed at three priorities:

- teaching about healthy transmission-prevention techniques
- caring for the millions of children orphaned by AIDS
- treating those already infected

Sub-Saharan Africa accounts for more than 30 million of the 40 million people with HIV/AIDS worldwide. It also accounts for 15 million of the estimated 20 million deaths caused by the disease. The continent is home to 11 million orphans whose parents have died from AIDS. While these statistics are truly horrific, the situation is not universally bleak. There are some rays of hope, thanks to a handful of responsible governments. For example, the government of Senegal showed foresight by launching prevention programmes at a time when the country was largely untouched by the disease. This seems to be paying off — only 27 000 of its 9.6 million people are HIV-positive. Uganda is another 'beacon of hope' country. In the early 1990s, the HIV prevalence rate reached 14%, with some urban areas reaching 31%. The national rate has now fallen to 6.1%. A concerted anti-HIV campaign, backed by the country's religious, traditional and civic leaders, bombarded Uganda's 23 million people with an anti-HIV message. This resulted in a substantial increase in sexual abstinence among the young and an increase in condom use from about 7% to 50% in rural areas and to 85% in urban areas.

Compare these two encouraging examples with the situation in Botswana, where there is a 39% infection rate in a population of 1.5 million. Even worse is the case of South Africa where, until very recently, President Mbeki was questioning whether HIV/AIDS existed at all! Yet South Africa has the largest HIV-positive population in the world (over 5 million in a population of 44 million). Roughly 40% of all adult deaths are now due to AIDS. By 2015, the disease is expected to have lowered the population by 12 million. Projections of present trends in Botswana are still more alarming. They raise the distinct possibility that its population could become extinct in a matter of decades unless some decisive action is taken immediately.

Figure 54 illustrates how the issue of the spread of HIV/AIDS touches all five components recognised as a part of population geography. However, the contact is a

Figure 54 The spread of HIV/AIDS: contributory factors and consequences

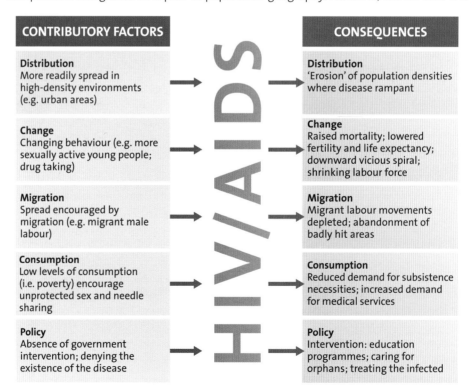

CONTRIBUTORY FACTORS	CONSEQUENCES
Distribution More readily spread in high-density environments (e.g. urban areas)	**Distribution** 'Erosion' of population densities where disease rampant
Change Changing behaviour (e.g. more sexually active young people; drug taking)	**Change** Raised mortality; lowered fertility and life expectancy; downward vicious spiral; shrinking labour force
Migration Spread encouraged by migration (e.g. migrant male labour)	**Migration** Migrant labour movements depleted; abandonment of badly hit areas
Consumption Low levels of consumption (i.e. poverty) encourage unprotected sex and needle sharing	**Consumption** Reduced demand for subsistence necessities; increased demand for medical services
Policy Absence of government intervention; denying the existence of the disease	**Policy** Intervention: education programmes; caring for orphans; treating the infected

double one. Two distinct contexts of cause (contributing to) and effect (consequences) are involved. The consequences are first and foremost demographic. The diagram also hints at some of the courses of action needed to fight the spread of the pandemic. For example, education is critical to the success of any anti-AIDS campaign. The young need to be warned of the dangers of sexual promiscuity, unprotected sex and needle sharing among drug addicts. Transmission of the disease is particularly encouraged by the migratory labour systems that characterise much of southern Africa. These involve men leaving their homes and families to find work, sometimes abroad or in distant cities, often in the mining industry and transport. Time away from home often leads to time being spent in brothels. However, there are also customs and practices deeply embedded in African culture that put people at risk, including:

- polygamy (taking more than one wife)
- inheriting the wife of a deceased brother
- communal breast-feeding, which is still widely practised in rural areas

It is beginning to be recognised that the spread of HIV/AIDS is a global issue:

- There are few, if any, countries where the disease is not encountered.
- The global community has a duty to help those countries most afflicted by AIDS. This obligation is heightened by the fact that many of those countries are poor. They simply do not have the resources and the organisational structures to halt the spread of the disease. Anti-retroviral drugs are now available; these need to be made freely available to sufferers, particularly the poor ones.
- There is the argument that until HIV/AIDS is curbed in sub-Saharan Africa, there remains the risk of it spreading to and re-infecting other parts of the world.
- The pandemic is worsening severely in countries that have so far escaped lightly, particularly the 'demographic giants' of India, China, Russia and Indonesia.

The potential demographic and wider consequences of the spread of HIV/AIDS are so horrifying as to be almost unthinkable. The only hope for the future lies in a combination of bottom-up and top-down strategies, ranging from more responsible individual behaviour and better education to governmental leadership and international aid.

Examination advice

The case studies that have been an integral part of your A-level geography course can be put to good use in a variety of ways in the examination. Much depends on:

- the question task and the command used in the wording of a question. Table 14 indicates that there are at least four different question scenarios: name, support, compare and examine. They are ranked in terms of what is expected in the degree of case study detail, which ranges from simple name-dropping to detailed knowledge and understanding of a particular situation.

- the context of the question. The challenge here is to find an example or case study that is appropriate to the question topic. Table 15 should help, because it links the 29 case studies presented in this book to the main topic areas of population geography.

Table 14
Question commands and the examiner's case study expectations

	Required case study detail			
Task or command*	Name	Support	Compare	Examine (a particular situation or statement)
Typical question	Give an example of a country at Stage 1 in the DTM	With the use of examples, explain why the relationship between population and resources is so important	Evaluate the birth control programmes of two named countries	With reference to a named country or region, examine the main factors affecting the distribution of population
Case study expectation	No more than the name of an appropriate country, e.g. Ethiopia, Bangladesh	Use of at least two contrasting case studies, e.g. an over-populated and an underpopulated country; more than name-dropping needed	Better to opt for two contrasting case studies, e.g. India (voluntary; limited success) and China (enforced; successful)	A sketch map showing the salient features of the distribution is a minimum requirement; a systematic look at key factors should follow

*A word of caution — watch out for those questions that do not specifically ask for examples, but nonetheless expect them. For example:

- Suggest reasons why some migrations involve hazardous journeys.
- 'Migration creates as many problems as it solves.' Discuss.
- To what extent do you agree with the view that the global population problem is not one of numbers, but of distribution?

If you are in doubt as to whether examples are required in your answer to a question, it is better to give some rather than none.

In the remaining sections, we look at five different examination contexts that require the use of case studies. There are the four shown in Table 14; they apply mainly to unseen examination papers. The fifth is where the geography specification requires you to undertake an enquiry into a set topic and to submit a report by a given deadline (e.g. Unit 5 of the Edexcel Geography B specification). Therefore, the contexts range from simply naming an example of a particular situation, through using case study material for support purposes, to an extended and detailed use of case studies.

Table 15
A matrix relating case studies, figures and tables in this book to the main topics of population geography

Topic	Case study, *Figure* or Table	Topic	Case study, *Figure* or Table
Distribution		**Migration**	
Using census data	1	Different types of migration	7, 8, 9, *18*
A systems view of population	*1, 8, 9*	Push and pull factors; motives	7, 9, 14, *18, 19, 20, 26*, 3
Portraying distribution, density *and change*	2, 3, *3, 6, 7, 29, 46*	Present patterns of international migrations	23, *19*, 44
Global patterns and related issues	21, 26, *10*, 6	The selectivity of migration	7, 9, *47*, 8, 9, 10, 11
Factors affecting distribution and density	2, 3, *2*	Benefits, costs and other consequences	7, 9, 18, 19, *23*
Changing distribution patterns	8, 25, 6	Controlling population movements	18, 25, 28
Managing distribution	14, 15, 27	Refugee and asylum seeker problems; ethnic cleansing	19, 23, 28
Change		**Consumption**	
Natural change, migration balance and net change	5, *8, 9, 48*	Population growth and resources	*27, 28, 32*, 12
Measures of population change and global variations	21, *10, 11*, 6	Indicators of well-being and quality of life	12, *5*, 12
Links between population change and development	4, 5, 6, *12, 14*	Coping with population growth	13, 26, 27
World population growth and future scenarios	21, 29, *42*	Overpopulation	11, 24, 27, *38*
Controlling population numbers	16, 17, *11*, 15	Underpopulation	11, 16, 20, 22
Changing age structures and gender roles	4, 5, 6, 20, 22, *12, 15*	Sustainable and optimum populations	15, 17, 20
Changing ethnic composition	5, 10, 25, *8, 9, 10, 11*	Impacts of consumption	12, 13, 24, *45*

Naming examples

This requires nothing more than being able to cite one of your case studies as a relevant example. The test here is one of appropriateness. In part (b) of the sample question below, it would *not* suffice to name just any densely populated country, such as the Netherlands or Singapore.

Question

(a) Identify three symptoms of overpopulation.

(b) Name one country that is widely recognised as being overpopulated. Give reasons for your choice.

Guidance

(a) Very high population densities; malnutrition or starvation; widespread poverty; high incidence of disease; environmental stress

(b) China/Indonesia/Puerto Rico/Haiti

The reasons are to do with the symptoms shown by the chosen country. Does it show all the symptoms, or only some?

Naming and using a supporting example

The next level up requires that you do more than just name an appropriate example. You are also expected to demonstrate knowledge and understanding by providing some detail.

Question

With reference to a named MEDC:

(a) identify a region of low population density

(b) outline the physical factors that help to explain that low density

(c) suggest one economic development that might help to raise population density

Guidance

(a) The island of Hokkaido (Japan)

(b) ■ Harsh climate, particularly in winter — inhospitable living conditions
   ■ Prevalence of mountainous terrain and thin soils
   ■ Remoteness from the 'core' of Japan

(c) The promotion of tourism — winter sports and wilderness adventure activities

Using case studies comparatively

Essay-type questions that require the comparative use of case studies are popular with chief examiners. *Using case studies 13* (page 52) has already provided one such example. Much of the challenge of such questions hinges on selecting appropriate case studies. In some instances, the choice is fairly obvious and restricted. In others, there may be more choice and options than you might first think from an initial reading of the question.

The trouble with all examination questions that ask you to compare situations (as represented by appropriate case studies) is that some candidates believe that all they have to do is to rehash each study in turn. This leaves examiners to draw their own conclusions as to whether or not the situations are similar. In short, the question is not answered and relatively few marks can be gained. In planning effective answers to the comparative type of question, it is necessary to interleave references to your chosen case studies. Look back at *Using case studies 13* (page 52) for an illustration of how this might be done.

Look back at *Using case studies 13* (page 52)

21 **Using case studies**

Question

With reference to two contrasting examples, evaluate the role of government policies in influencing rates of population change.

Guidance

The obvious choice here is to compare a country that has tried to curb its birth rate with one that has done the opposite. Romania (*Case study 16*) and China (*Case study 17*) are two immediate candidates. Equally, you would be justified in choosing China and India as examples of government intervention with two different outcomes — success and relative failure respectively.

Take a closer look at the question. There is nothing here to say that your answer can only be about policies to do with birth rates. Any country that has positive policies about health and medical services stands to affect rates of population change through increased life expectancy and reduced mortality rates. So, you could build an answer around the comparison of the UK (*Case study 5*) and Russia (*Case study 6*).

When you take this closer look at the question, you will perhaps begin to realise that there is nothing in the wording to say that you should not focus your answer on migration rather than natural change. A discussion involving a comparison of an 'open-door' country such as Spain (*Case study 8*) or Australia (*Case study 25*) today with a 'restricted entry' country such as the UK would allow you to pick out some contrasting demographic changes with respect to birth and death rates, age structure and ethnicity.

Building an essay around a single case study

A successful attempt to answer any question that starts 'With reference to a named country…' requires three things:

- choice of an appropriate country
- sufficiently detailed knowledge of that country in relation to the particular question
- resisting the temptation to set down all that you know about your chosen country (the 'everything but the kitchen-sink' approach), but instead harnessing only those aspects that are directly relevant

22 **Using case studies**

Question

For a named country, examine the demographic, economic and social consequences of a declining fertility rate.

Guidance

Given the case studies presented in this book, there is much to commend the UK as your choice (*Case studies 1* and *5*). You might find some prompts in *Case study 22* about the sorts of consequences that should be covered.

Demographic consequences include:
- a fall in the rate of population growth
- 'greying' of the population structure
- a shift in the nature of dependency from the young to the elderly
- encouragement of immigration to make good the shortfall in births

Economic consequences include:
- falling unemployment and labour shortages
- rising per capita tax burden to pay for services increasingly oriented towards the elderly
- changing consumer demand
- a shift in service needs — more residential homes for the elderly; fewer schools

Social consequences include:
- a decline of youth culture
- changing values in a 'greying' society
- innovation giving way to experience
- fewer mothers and more career women

Planning an extended essay involving a range of case studies

The standard advice on essay planning should be followed, namely that the essay should have a three-part structure — a brief **introduction** and an equally brief **conclusion** (one paragraph each) separated by a series of paragraphs (the **expansion**) that develop your argument or discussion points. It is here that you should incorporate supporting examples and case studies.

23

Question

Which do you think poses the greater challenge to a country — a high rate of natural increase or a large influx of immigrants? Justify your viewpoint with reference to examples.

Guidance

Introduction

Rather than jumping in at the deep end and naming your choice, you might be a little more subtle. You could start by making the point that since both scenarios are likely to result in rapid population growth, they therefore pose the same broad challenges.

Expansion

First, look at those common challenges. They include more mouths to feed, increased pressure on resources, need for more housing, raised demand for services, and the spatial distribution of the resulting population growth.

Now move on to pose the question in a slightly different form — given these similarities in outcome, are there any differences? If so, are these differences significant when it comes to government policies aimed at controlling each of the two situations?

Now look at the outcomes and challenges that are specific to each of the two scenarios. With respect to natural increase, there will be a need for more child-oriented services, such as clinics, midwifery, schools, etc. Contrast an LEDC (e.g. the Gambia or French Guiana), where the resources simply do not exist to meet this particular demand, with say one of the oil-rich states (e.g. Kuwait or UAE). With respect to immigration, outcomes and challenges include:

- immigrant characteristics (e.g. ethnicity, skills)
- attitudes and perceptions of the native population
- integration into society through education and equal opportunities

Once again, the UK would serve well as an example (*Case studies 10* and *28*).

Conclusion

Large-scale immigration poses more immediate challenges. However, a high rate of natural increase could be more threatening: first, because most countries experiencing it rank among the least developed; second, because the time-lag between taking action (i.e. introducing a birth control programme) and seeing any significant impact on the rate of natural increase is likely to be a long one. In contrast, it is possible to close the 'immigration door' fairly swiftly.

Grasping and applying the advice contained in this part of the book should have a positive outcome. It is sometimes easy to forget that geography is about the real world. The more contemporary examples that you can include in your examination work, the more you are likely to convince and impress the examiner that you have a sound knowledge and understanding of today's world. A good dose of reality in the form of relevant case studies and examples can work wonders when it comes to raising AS and A2 geography grades.

Appendix

Finding and organising case studies

This book contains 29 case studies that can be used to illustrate and support the key ideas relating to population and migration. However, you may choose to find some of your own. To do this, you might tap into the following sources:

- A-level textbooks — where the case studies are specially chosen
- magazines and periodicals — such as *Geography Review*, *Geofile*, *Geo Factsheet* and *Geo-News Review*
- videos/CD-ROMS and television programmes — you will need to take notes; the latter have the advantage of being up to date
- the internet — the opportunities are overwhelming; you will need to restrain your searches. Websites that currently provide useful information and material about population and migration include:

  www.demographia.com
  www.neighbourhood.statistics.gov.uk
  www.populationconcern.org.uk
  www.statistics.gov.uk
  www.unAIDS.org
  www.worldrefugee.com

- your local area — local newspaper, local authority plans and statistics

Case study data are best kept in a succinct note form, perhaps on filing cards, with key facts and figures set out as bullet points. Alternatively, much can be summarised by annotating a sketch map or diagram.

As you collect your case studies, you should cross-check them against the content of your AS and A2 geography specification. You need to ensure that you have good and even coverage of:

- the key ideas
- countries and places at different stages of development and at different spatial scales

You should also be aware that some case studies could be used to support more than one of the key ideas.